Management by Process

Management by Process

A Roadmap to Sustainable Business Process Management

By

John Jeston and Johan Nelis

AMSTERDAM • BOSTON • HEIDELBERG • LONDON • NEW YORK • OXFORD
PARIS • SAN DIEGO • SAN FRANCISCO • SINGAPORE • SYDNEY • TOKYO
Butterworth-Heinemann is an imprint of Elsevier

Butterworth-Heinemann is an imprint of Elsevier
Linacre House, Jordan Hill, Oxford OX2 8DP, UK
30 Corporate Drive, Suite 400, Burlington, MA 01803, USA

First edition 2008

British Library Cataloguing in Publication Data
A catalogue record for this book is available from the British Library

Library of Congress Cataloging-in-Publicaion Data
A catalogue record for this book is available from the Library of Congress

ISBN–13: 978-0-7506-8761-4

For information on all Butterworth-Heinemann publications visit our website at books.elsevier.com

Typeset by Charon Tec Ltd (A Macmillan Company), Chennai, India
www.charontec.com

Printed and bound in Hungary

08 09 10 10 9 8 7 6 5 4 3

Dedication

We would like to thank our families (Yvonne, Brittany, Connor, Cassie & Kurt and Sandra, Angelique & Mystique) for providing us with the time and opportunity of writing this book.

John and Johan

Table of Contents

Contributors

John Jeston has been working in business and IT for over 30 years, covering business process management, business process re-engineering, project management, systems development, outsourcing, consulting and general management. In addition to his consulting roles he has held the positions of financial controller, divisional manager, director of a software company, HR director and CIO.

John now specializes in business process management, helping organizations by providing BPM training, coaching and a framework to enable operations management to become high performance managers. He advises organizations in strategy, provides coaching and training, consulting advice with regard to performance- and process-based project implementations. John is the managing partner of an independent specialist consultanting practice, *Management by Process Pty Ltd* (www.managementbyprocess.com). John, his partners and team build internal BPM skills and capabilities within organizations; complete Board and CEO business process reviews; provide clients with a 'road map' for the creation of a sustainable, successful and repeatable business process improvement programmes and process-based change management

He presents at conferences, works with organizations and its leadership in promoting a process-focused high performance management environment, holds workshops, and is the co-author of the highly successful publication: *Business Process Management: Practical Guidelines to Successful Implementations* (2006 and 2008). He is also an author and course director of a BPM distance learning programme in Australia and writes a regular column for BPTrends. John can be contacted at johnjeston@managementbyprocess.com.

Johan Nelis has international experience as a hands-on Business Process Management Consultant. He has established and managed a BPM practice of 30 consultants in The Netherlands, and also co-founder and Vice Chairman of the Dutch BPM Forum. Johan has worked for the United Nations as an Advisor. He is well known for his drive to transfer knowledge and experiences, and has shown that he is capable of motivating and stimulating people. He has initiated many BPM training courses and lectured at a postgraduate course.

Johan has carried out assignments in a wide variety of sectors, with the main emphasis on finance and telecom. He specializes in aligning processes with strategy, business objectives and IT. He has also performed many process audits, and is able to pinpoint fundamental problems, formulate quick wins, and provide innovative and sustainable solutions. Furthermore, he is good at initiating and overseeing implementations of BPM and ensuring that the people are able to perform their activities better and independently. Johan is a lead consultant at a specialist BPM consultancy where he provides strategic advice on business process services and supervises a team of BPM consultants. He has presented at seminars and hosted workshops at several BPM conferences in Europe and Australia. Johan is an author and course director of BPM distance learning programmes in The Netherlands and Australia and writes a regular column for BPTrends. Johan can be contacted at johannelis@managementbyprocess.com.

Mandy Holloway's continuing journey from emerging leader to partner at KPMG and for the last six years developing business leaders, while juggling the roles of wife and mother provides the incredible platform of realism she brings to any leadership development initiative she delivers for her clients.

She brings a passionate and energetic focus that unleashes the courage within each person to be the kind of leader they want to be and the leader the business needs them to be while they negotiate the rapid changes imposed everyday!

Emerging leaders and existing leaders have their thinking challenged and their hearts engaged at a very real level when Mandy works with them. She nurtures them to build personal confidence so their courage grows and they gain the personal conviction to make the right choices for themselves and for the business.

The outcome is authentic leaders generating sustainable personal and business performance.

Dr Tony Gardiner has over 16 years of consulting and commercial experience in business process improvement at the strategic and operational levels. He is currently a Process Executive at Nedbank in South Africa focusing on service modeling and business process maturity. Previously Tony worked as a management consultant leading business analysis teams, business process re-engineering and organizational development efforts across a wide variety of industries. At Nedbank he has led process integration efforts across the full retail process spectrum, as well as the re-engineering of the loans business processes. Most recently he has led the development of the Service Operations Capability Maturity Model, now the Business Process Maturity Model (recently adopted as an international standard by the OMG), and the development and institutionalisation of the supporting process solutions. He has a BSc (Zoology), BSc Hons., MBA and PhD all from the University of Cape Town.

Foreword

Jeston and Nelis have given us a highly reasonable approach to the advocacy and implementation of process management. They not only simply and calmly lay out the principles of managing by process, but also present several in-depth case studies about organizations that have realized how important process management is to their success. If process management works for a large, global bank such as Citibank, manufacturing organizations like Air Products, and medium-sized health care organizations like Aveant Home Care, why wouldn't it work for your organization? The answer is that it probably would.

The other appealing aspect of this book – true also of the authors' last book, *Business Process Management: Practical Guidelines to Successful Implementations* – is that they are agnostic as to which particular approach to improving processes you should employ. As they note in the first chapter, continuous improvement is well-suited for some leading firms, but not if you need rapid and radical improvement in your processes. This seems common-sensical, but it is all too rare in the process management world for leading thinkers to approve of multiple different approaches to process change. The insistence on a particular approach – be it Six Sigma, Lean, TQM, reengineering, or whatever your favorite – has probably been one of the reasons why process management in general has not developed as it should. No particular approach to process management encompasses all of the methods, tools, and objectives that a large organization needs in managing its processes. Hence the synthetic, agnostic approach taken in this book is almost always best.

One last process management crime of which this book and these authors are not guilty is over-engineering. Advocates of process management often believe that the world presents a great opportunity to be engineered and re-engineered. These people believe that a detailed process flow diagram is the answer to virtually every problem of organizational performance. Jeston and Nelis are not members of the over-engineering fraternity. They realize that organizations and their processes are comprised of people, and that process flows and maps – while undeniably useful – are only plans for how work should be done. As with any sort of plan, getting people to actually follow a process is a matter of leadership, change management, and human culture and behavior. Unlike many books on process management, in this volume you'll find as much – probably more – content on human change issues than on the engineering aspects of processes.

As the authors note, process management isn't a silver bullet – but it is a bullet. It's probably the best way to get lasting improvements in operational performance from your organization. It's the best way to reduce variation in how work is done, and to surprise and delight your customers with your consistent meeting of their expectations. It's the best way to reduce unnecessary costs and time as you do your work. Read this book, implement these ideas, and you will be on your way to achieving these long-sought yet entirely practical goals.

Thomas H. Davenport

Preface

After the success of our last book *Business Process Management: Practical Guidelines to Successful Implementations* we believed we had only told part of the process story. We had provided readers with a proven, successful and repeatable framework for the implementation of BPM projects and shown that these projects need to be handed over to the business for the sustainability of the business processes improved as part of the project.

We also provided a chapter called *Embedding BPM within the organization*, but this only briefly touched the surface of what is required by organizations to become business process-focused and have high performance management.

When speaking to, and consulting with, organizations on how to create an awareness of business processes; how to have managers and staff become more process-focused and managed; we have found that they often do not know where to start or what they should be aiming to achieve. Organizations realize that they have a long way to go to achieve an ideal state but have no overall structured approach of how to get there and what steps they need to take.

This journey is a large and complicated set of tasks for an organization, so we have broken the journey down into seven dimensions. We have then provided, for each dimension, an example of the ideal or, as we have called it, a visionary state and then provide a practical roadmap of how to get there.

Furthermore, we have provided a number of lengthy and robust case studies of successful BPM implementations. Most organizations have provided permission to be identified, but not all. We compiled these case studies by either interviewing the people who completed the work within the organizations and asked that they tell their story; and in one case, be the author of the case study. In one case study we told the story of an engagement with a client.

John Jeston and Johan Nelis

Preface

Acknowledgements

Our aim is to always provide practical advice to our readers based on real life experiences from implementing and consulting with organizations – this book is no different. Many of the experiences and ideas have come from our work with clients, and the issues and problems they face.

While we have valued the advice and contribution of friends and colleagues, ultimately we must take responsibility for what is in this book.

Without the reviews, contributions, critical comment and robust debates with these friends and colleagues, this book would not have been possible.

As always there are a few special people who we would especially like to thank.

Guus Balkema from YNNO consulting in The Netherlands for his review and comments on drafts of the book; Mandy Holloway for her robust debates around leadership and contributing to the writing of the 'Process leadership' chapter; Tony Gardiner for his contribution in writing the Nedbank case study; Michel van Drie; George Diehl for his wonderful discussions and interview for the Air Products case study; Saju Madhaven and Manuel Loos for their review and contribution to the Citibank case study; Evert Mulder, the inspiring Chief Executive Officer of Aveant for his case study; and Antje Breer for her robust discussions around functional versus process focused organization structures.

We would also like to thank Mike Overwater and Ken Beattie for their discussion and contributions to the High Performance Management model.

Finally, our great thanks to our senior editor, Maggie Smith, for her trust, support, patience and good humor throughout this journey.

How to read this book

In writing this book we have described the management of an organization from a business process perspective. This requires an examination of the execution and management of end-to-end business processes and all the aspects that are needed to support them. When managed well, it will enable an organization to move from its current position towards becoming a high performance management business.

Processes permeate all aspects of a business and they are increasingly going beyond the boundaries of individual organizational functional silo's and even the organization itself. This means that these organizational functional silo's and external parties must find ways of collaborating.

An example of the need for this collaboration is the emergence of the shared service environment. These are where common service areas of a business (often back-office functions) are centralized and 'shared' across other parts of the organization. These shared services areas can be created within an organization and/or outsourced or off-shored to other organizations. Unless handled well, the interfaces between the organizations and its processes can be difficult and actually make the service levels worse.

The increased connectivity of technology provides opportunities for organizations to have work performed in a more seamless manner. For example, document and workflow management allows the back-office activities to be performed off-shore; or the use of web-services allows the credit checks to be performed instantly.

Most organizations were designed for the industrial age of the past century, when capital was the scare resource, interaction costs were high and hierarchical authority and vertical integration structures were the key to efficient operation. Today superior performance flows from the ability to fit these structures into the present century's very different sources of wealth creation. (The McKinsey Quarterly, 2007)

There is a need to remake the organizations of today into more adaptive, agile and focused organizations and this is what we have attempted to show in this book.

There are three parts to this book. Part I provides a brief overview of the importance of business processes within an organization. We then provide six robust and lengthy case studies of organizations that have taken, or are taking, the process journey and their successes and failures. The case studies may be read before or after Part II, this is your choice. You may find it useful to quickly read them, then read Part II in detail and revisit the case studies again to learn deeper lessons from them.

Part II is about the Management by Process framework that will allow an organization to become a process-focused organization and attain a high performance management environment. We provide an introduction to the framework by discussing organizations from both a functional and process viewpoint; an outline of the seven dimensions of the Management by Process framework; and then discuss each dimension in detail. During this detailed discussion, we will provide an explanation of why each dimension is important; what are the key trends associated with it; what are the key elements; describe the visionary state (what should an organization aspire to); and finally we provide a roadmap of steps required to attain a sustainable process-focus and high performance environment. If an organization wishes to sustain its process improvement and management effort, then this is a roadmap that works.

Part III is the appendices which provide additional detailed information about several of the dimensions. This comprises information we consider useful, but too detailed for Part II.

The book also contains a number of smaller case studies throughout to illustrate various points in the book.

Part I

Management overview

In this part of the book we discuss two aspects from a management view. Firstly, we briefly discuss the importance of business processes to an organization. We then go on to demonstrate, via a number of detailed case studies, how various organizations around the world have taken a business process-focused perspective and dramatically benefitted from it.

These case studies are from Europe, South Africa, Australia and the USA and include large and prestigious organizations such as Citibank, Nedbank, Aveant Home Care services and Air Products and Chemicals Inc. All but two of the organizations has provided us with permission to use its name.

While you may find some of the case studies long and detailed, we felt it was better to provide you with all the detail than not enough. You can always scan them and come back and read them in detail at your leisure. We would however draw your attention to one aspect when you read the cases – the role of the organizations leader in the process-focused journey.

Chapter 1

Introduction

In the introduction we would like to address four topics which focus on the broad context of business processes and how they impact upon the performance of the organization. The topics are:

1. The importance of business processes.
2. A brief discussion of Systems Thinking and why this is important in the context of business processes.
3. The Management System within an organization proposed by Kaplan and Norton in their Harvard Business Review article (January, 2008).
4. Characteristics of a High Performance Organization.

In discussing the topics, we will provide a brief introduction and then provide more detail in Chapter 3 and map the suggestions to the framework proposed in this book for the sustainability of business process management.

Importance of business processes

Most executives struggle with the concept of why business processes are important to an organization. Let's face it historically there has been little formal tertiary management education on the opportunities that business processes bring to an organization or the impact on an organization if they are sub-optimal.

Some of the recent literature in the process world has suggested that business processes are so important that the organization structure should be turned upside down to be a process-centric organization, rather than functionally based. It is argued that changing from the traditional functional, hierarchical orientation to a process-centric orientation will mean that our organizations will function with greater efficiency and effectiveness, to the benefit of management, staff, customers and all other stakeholders.

After all, a functional organizational structured view creates a silo effect within an organization, and this often leads to selfish or self-centred behavior by the management and staff of each silo, sometimes to the detriment of other silo's and the organization as a whole.

Most organizations continually complain about the impact of organizational silos and the harm it is having upon its business. There is often significant effort expended attempting to minimize, or eliminate, this silo effect but it can take years and years to orientate all the management to a more holistic approach and behavior. If successful, the challenge then is to maintain this new found focus as the management and staff come and go from the organization. If this is not successfully passed from one manager to another, then the organization can regress back again to a silo'ed situation. After all, this is how business has successfully functioned for decades.

While a process-centric structured organization can, in certain circumstances significantly benefit an organization, is this always true?

We wonder if the same people who stand up at business process management (BPM) conferences singing the praises and necessity of having a totally process-centric organizational structure will, several years after its adoption, be complaining equally and vigorously about this structure, in a similar manner to the way they currently complain about functionally based organizations.

Even if an organization achieved the perfect organizational structure, this is still not a 'silver bullet' for the future success of an organization. Business organizations are complex and intertwined organisms with no one aspect being dominant or the 'silver bullet' to solve all its challenges and issues.

The continual and sustainable success of an organization is a complex set of interacting events and criteria and much has been written on how to achieve synergy.

Results are driven by many things, but at a high level, it is the organizations leadership that provides the vision, strategy, targets, organizational structure and operational efficiencies to achieve the strategy and objectives.

We would argue that it is predominately business processes that provide an organization's ability to deliver products and services to customers. Business processes are the link between all aspects of an organization. Processes are the link between an organization and its:

- suppliers
- partners
- distribution channels
- products and services
- people (personnel)
- other stakeholders.

Therefore we see business processes as the central core from which business is conducted, so long as they are supported by the people within the organization.

Organizations exist to supply customers with products and services, and business processes are the means via which this can be achieved to, hopefully, a high level of service and satisfaction. So customers must be the primary focus for business processes. Ultimately, BPM is a community of people working together with a common goal of providing a solution, product or service to customers ensuring they are serviced to a high standard and leave the experience delighted.

It is critical to also say at this point that we do not see *process improvement* as a silver bullet or the *only* 'game in town' when it comes to achieving results within an organization. While process improvement or redesign can, and will, make a significant difference to an organization, it cannot achieve the results required by an organizations strategy without an organization's senior management activating other aspects or components that comprise an organization and lead it to success.

An organization may have the best, most efficient and effective, business processes in the world, but unless they have a product or service that customers desire and demand, then 'who cares'.

Having said this, having sensational business processes will provide an opportunity for what Porter refers to as *competitive advantage* (Porter, 1980). This is where an organization manages to dominate an industry for a sustained period of time. Porter goes further and suggests that organizations obtain a competitive advantage either by an 'operational effectiveness' approach or a 'strategic positioning' approach.

What is the difference between these approaches and why does it matter?

Operational effectiveness approach

This is achieved by an organization creating 'best practice' within their organization. High achievement means having, or adopting, a best practice approach. But best practice within one organization is not necessarily sustainable because organizations will 'just copy you', or at the very least, apply considerable resources to obtain an equally effective operation.

According to Porter if one organization obtains a lead in a competitive marketplace, competitors will move very quickly to match them, especially if that lead has been established via the implementation of a new software application or new business processes.

The difficulty with this approach is that it is premised upon the fact that, in order to be successful, you will need to take market share away from a competitor, and they will not stand by and allow you to do this without a fight. With this type of competition, this approach has the potential of forcing organizations to invest more and more into efficiency gains and therefore driving down profits. This is the 'red ocean strategy' described by Kim and Mauborgne (2005).

Strategic positioning approach

An alternative to the above position is to create a unique position within your marketplace that will make it difficult, or ideally impossible, for others to compete. While Porter has called this strategic positioning, of recent times Kim and Mauborgne (2005) refer to this as a *Blue Ocean Strategy*. The latter argue that an organization should position itself into an untapped marketplace, thus changing the rules of the game being played. Both Porter and Kim and Mauborgne argue that redesigning or positioning an organization's *value chain* offering is an essential component in this strategic positioning process.

Customers and investors are looking for organizations to take a leadership position, to show vision and not just to improve on the past.

It is interesting that in today's business environment *constant improvement is not enough* – it is not good enough for us to be better than we were last year or better than our competitors.

According to Paul O'Neill, Chairman of Alcoa 1991, in a worldwide letter to Alcoa staff in November 1991:

> Continuous improvement is exactly the right idea if you are the world leader in everything you do. It is a terrible idea if you are lagging the world leadership benchmark. It is a disastrous idea if you are far behind world standard – in which case you may need rapid quantum-leap improvement.

The following is an *overriding brief* that we were given by one of our clients:

> We do not wish to implement an incremental change program, but a change program that will place us substantially ahead of our competition, such that it will be difficult for competitors to match the process and systems service levels able to be consistently achieved. This will form the foundation of our competitive advantage in the near and medium term.

The client wished to not only create a leap-frog effect that would place them substantially ahead of its competitors now, but to build an environment to support the continuity of radical change within the organization to maintain a competitive advantage.

This is not achieved by completing process improvement projects. It can only be achieved by a change in focus within the organization and this change in focus may only be commenced by a *program* of activities that *encompasses the entire organization.*

Systems thinking

We are strong believers in Systems Thinking because it provides the basis for a structured and consistent way of thinking and managing, and yet, allows for creativity. Creativity must always be built into the system and *ad-hoc* decisions can be taken when the need arises.

Systems thinking should happen at all levels of the organization: at the strategic and operational level as well as the interaction between them. The Deming circle (Walton, 1986) of Plan – Do – Check – Act, is an example of this. When you take into account the fact that some business processes can take days, weeks or months to complete, it is important to proactively monitor progress, so we have added a Monitor step. Refer to Figure 1.1.

Systems thinking would suggest that Management creates a Plan of what they would like to Do. This goes into the Execution mode that may either resemble a process, a project or smaller set of activities. The outcome of the Do step needs to be Checked and/or Monitored over time and at the completion of the activity. As a result of the outcomes, there will be triggered a need to take action (Act). This action will either resemble Process Improvement (if the action requires a change to the process, a process improvement project) or go back to the Plan step if no changes to the process are required (for example if more effort is required because of higher demand).

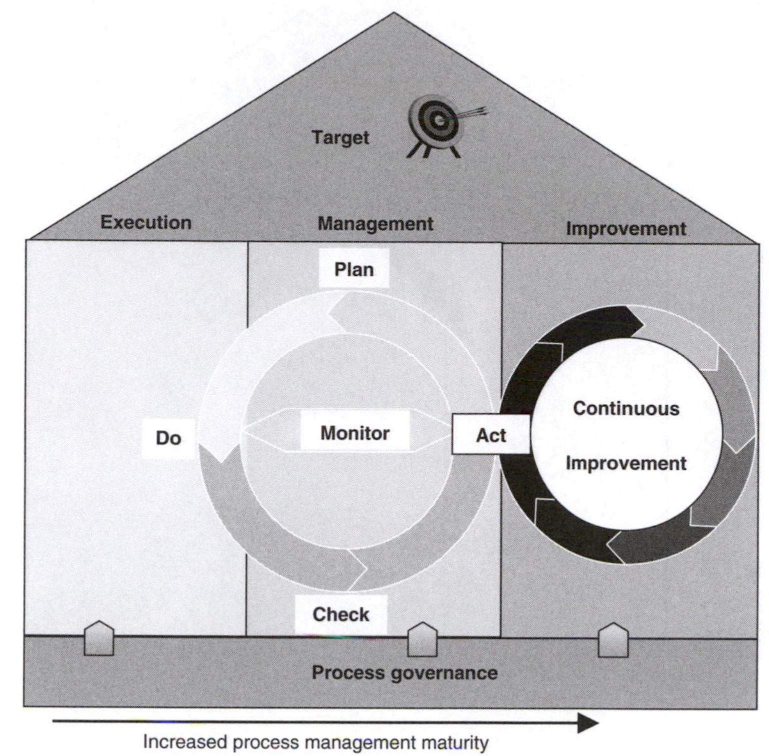

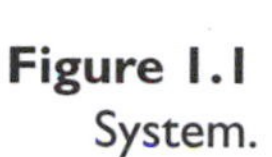
Figure 1.1
System.

We will now describe four situations that make the achievement of systems thinking and acting difficult or impossible to achieve.

1 *You cannot achieve your target, unless you manage it*
Targets and goals are rarely met without the involvement of "management" and management action. If the targets are met without management involvement, then they simply were not ambitious enough. Management provides guidance and ensures that the various pieces of the puzzle fit together (Figure 1.2). Management requires clear definitions of roles and responsibilities, including ownership.

2 *You cannot manage what you do not measure*
Management requires measurement. While the popular "management by walking around" is an important tool to gain a sense of what is happening "on the workshop floor", it can never be the only tool, nor replace true measurement of process and people performance. Refer the Figure 1.3 which shows a disconnect between the Do and Act steps.

3 *You cannot improve without management*
There are still many organizations that have a low level of business process management maturity and yet still attempt to start business process improvement activities without firstly establishing the management required for these processes. Even if the organization does achieve some process improvement, the gains will rapidly disappear unless the business processes are managed for sustainability. In our

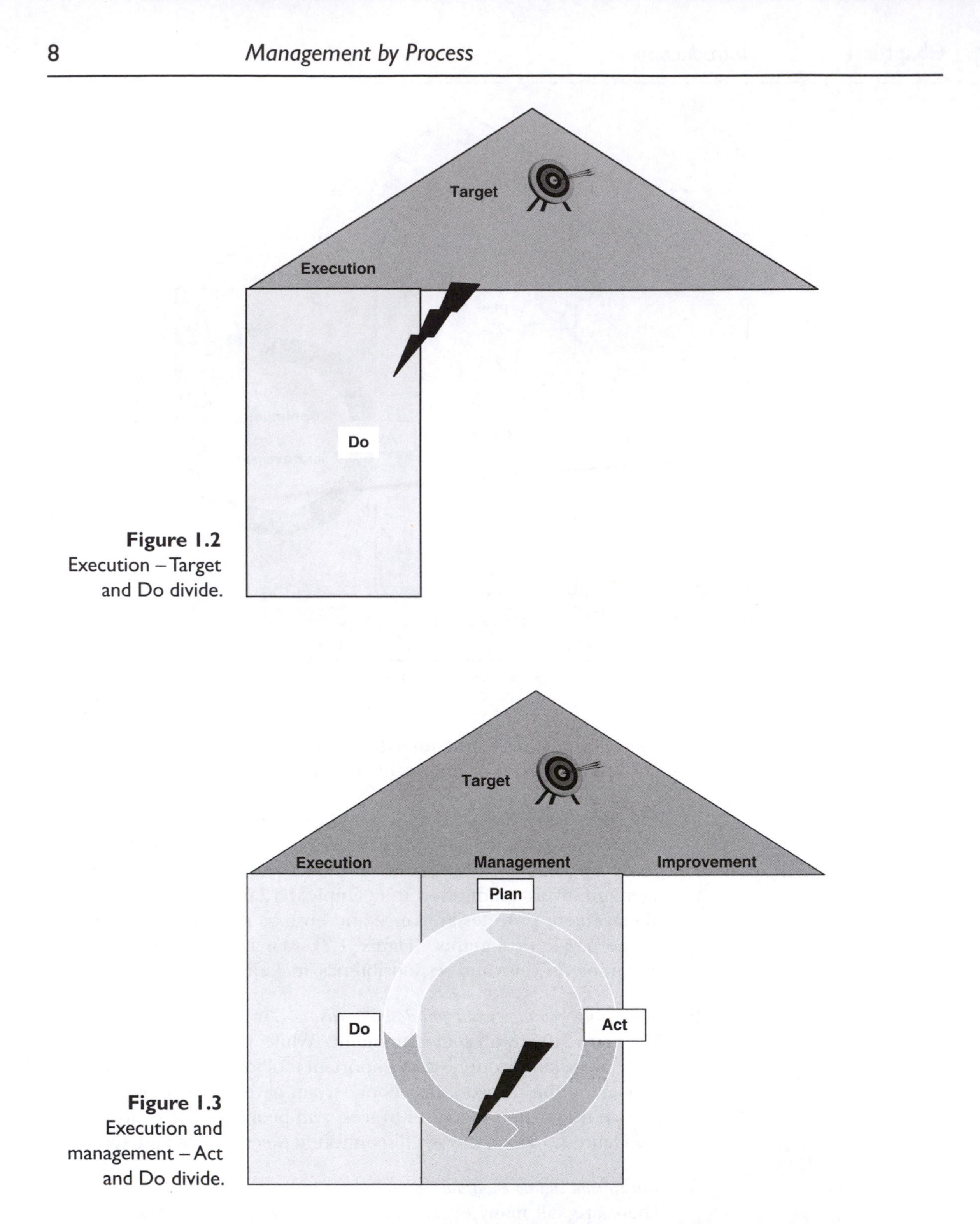

Figure 1.2 Execution – Target and Do divide.

Figure 1.3 Execution and management – Act and Do divide.

experience, many Six Sigma projects fall into this category. Figure 1.4 shows that unless the divide between the Do and Improvement steps is filled with the management of the activities to achieve the Targets, then the benefits associated with the improvements will simply diminish over time.

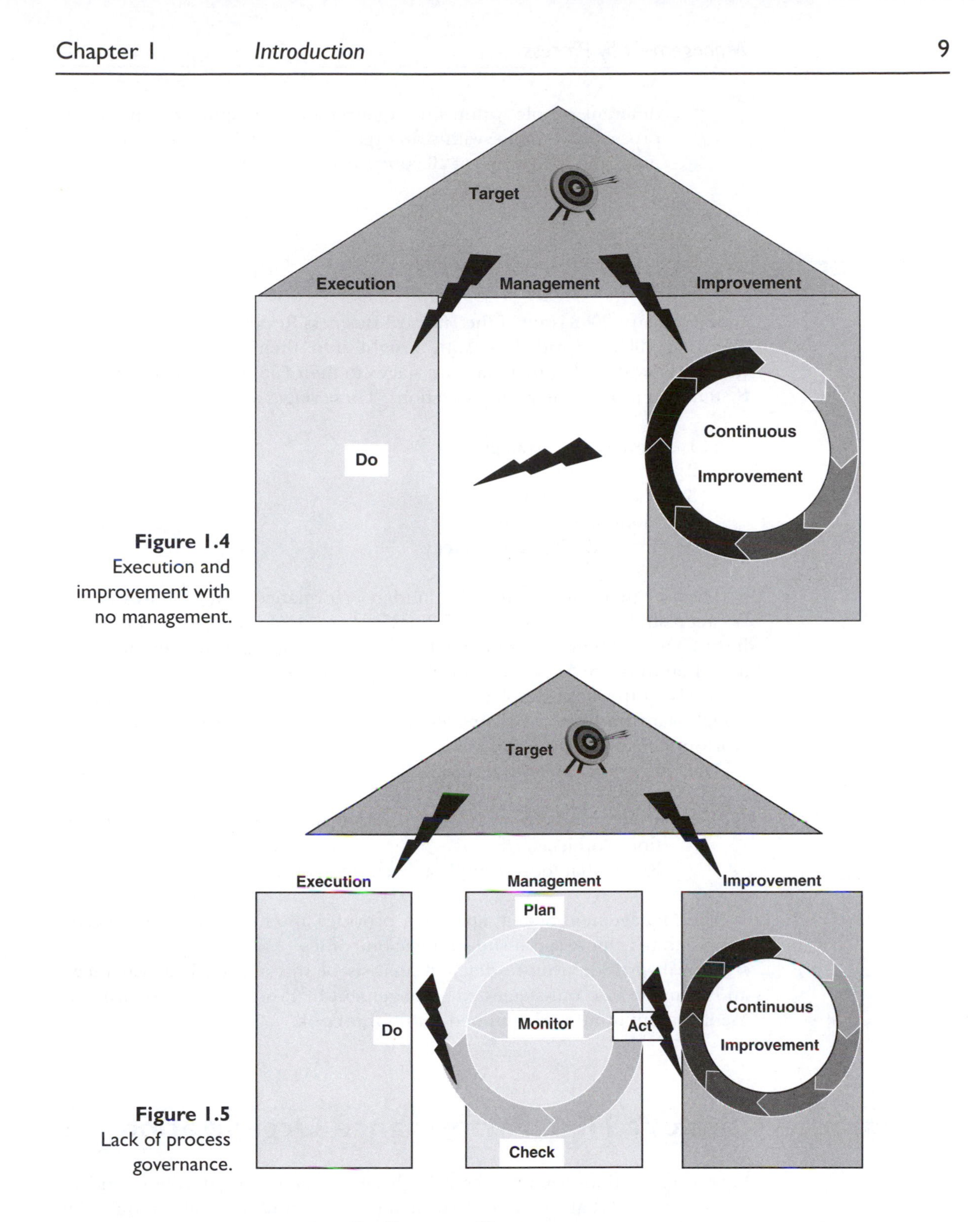

Figure 1.4 Execution and improvement with no management.

Figure 1.5 Lack of process governance.

4 *No alignment without governance*
Process governance, in our opinion, is the foundation of the systems thinking figure. Process governance must ensure that the target, execution, management and improvement activities are aligned. This is crucial as the various roles for these aspects are distributed among

different people within the organization. A pragmatic approach to process governance within an organization will increase the commitment and adherence of all concerned. Refer to Figure 1.5.

Mastering the Management System

In the January, 2008 issue of the Harvard Business Review, Kaplan and Norton (January, 2008) provided us with insight into their Management System model. They state that there are five stages to their Closed-Loop Management System that links strategy and operations. These stages are:

1. Develop the strategy
2. Translate the strategy
3. Plan operations
4. Monitor and learn
5. Test and adapt the strategy.

They say that "most companies' underperformance is due to breakdowns between strategy and operations." These steps, "describes how to forge tight links between them… A company begins by developing a strategy statement and then translates it into the specific objectives and initiatives of a strategic plan. Using the strategic plan as a guide, the company maps out the operational plans and resources needed to achieve its objectives" (Kaplan and Norton, January, 2008, p. 65).

They underline two critical success factors for successful strategy execution:

1. Understand the management cycle that links strategy and operations (projects)
2. Know what tools to apply at each stage of the cycle.

The Management System approach provides an enterprise encompassing approach to achieve successful strategic outcomes.

We will provide a more detailed analysis of this approach in Chapter 3 and compare it to our suggested Management by Process High Performance Management framework outlined within this book.

Characteristics of a High Performance Organization

Most organizations aspire to be a High Performance Organization, and yet few achieve it. What is a High Performance Organization, what is required to achieve it and indeed, why do we need to be one?

In 2005 Dr Andre de Waal published a paper entitled *The Foundations of Nirvana.* He and his team build on some other research they had completed earlier, and examined 91 High Performance Organization studies which had been performed in the last 15 years. They were interested in gaining an understanding of the characteristics of such organization.

The need for a High Performance Organization is underlined by the following facts (p. 5):

- More shareholder value has been destroyed in the last five years as a result of mismanagement, wrong decisions, and bad execution of strategy than was lost through all the recent compliance scandals combined. In a recent Booz Allen Hamilton survey among 1,200 large companies, it turned out that at the 360 worst performers 87 percent of the value destroyed, was caused by strategic missteps and operational ineffectiveness. Only 14% could be attributed to compliance failures or poor oversight of the company's corporate boards (Kocourek, 2005).
- The average time a CEO or managing director spends in the top-position is continually decreasing, from an average of more than ten years two decades ago, to two and a half years today (Tijdschrift Controlling, 2005).
- More than 50 percent of managers make decisions based on their gut feeling, not on hard facts and 36 percent have black boxes in the organization of which they know hardly anything (SAS Institute Nederland, 2002).
- Return rates and warranty costs are dramatically rising while at the same time customer satisfaction levels are steadily decreasing, a strong indication of the deteriorating quality of products (Kleiner, 2005).
- Of recent mergers and acquisitions, only 17% were reported to add value to the combined company, 30% produced no discernible difference, and 53% actually destroyed value (KPMG, 1999).
- The majority of companies that get into a crisis find themselves in this situation because of internal factors, of which dysfunctional management (48% of the cases) and inadequate management information systems (42%) are the most common causes (Eyck van Heslinga, 2002).

The definition that he formed was based on the common themes in the research examined, and it was proposed that a High Performance Organization (HPO) is:

> "an organization that achieves results that are better than those of its peer group over a longer period of time, by being able to adapt well to changes and react to these quickly, by managing for the long term, by setting up an integrated and aligned management structure, by continuously improving its core capabilities, and by truly treating the employees as its main asset." (p. 12)

It was suggested that "the management processes of an HPO are integrated and the strategy, structure, processes and people are aligned throughout the organization" (p. 12) and it continually improves and reinventing its core capabilities.

The study aimed to identify common characteristics among 91 studies in the area of HPO, focusing on actual and quantifiable research. He deployed the following criteria:

- Study must have been after 1990, so that it focused on the current age of extreme competition.
- Study consisted of either a survey with a sufficient number of respondents to be (fairly) representative or in-depth case studies.
- Written documents containing an account and justification of the research method and research approach.

The analysis concluded that there were 8 factors that influenced high performance, these were:

- External environment
- Organizational structure:

  – Organizational design
  – Strategy
  – Processes
  – Technology

- Organizations culture:
  – Leadership
  – Individual & Roles
  – Culture

The paper then classified the 91 studies into elements that relate to one of these factors. For each of the characteristics the "weighted importance" is calculated, that is, how many times it was mentioned in the various studies. A HPO characteristic required a weighted average of 5% and in at least 5% of the studies, bringing the total of characteristics listed to 68. For the purpose of this book we only present the top 20 characteristics. In Chapter 3 we will provide a listing of these top 20 characteristics and then relate them to our Management by Process High Performance Management framework.

The reason for discussing this is to clearly make the point that organizations are complex entities as are the management theories that support them and working only on one part of it alone, like business processes, will *not* solve all its problems nor guarantee success. We are simply saying that there are many other things to do than just redesign business processes to be more efficient and effective.

However, if an organization chooses to ignore its business processes, or not have a goal of moving towards being a process aware organization, then it will certainly not achieve a high level of sustainable results.

Before we move on, we would like to make it clear that this book is not about how you solve *all* these complex aspects within an organization. While we will concentrate on the business process side of the equation, this cannot be tackled in isolation of the other components. So while we address the business processes, we will bring in the other aspects as necessary, but only to the extent that they impact an organizations ability of becoming a process-focused organization.

If an organization wishes to make the move towards being more process-focused, then how can this be achieved? What are the activities that they must complete? What are the attitudes that must be influenced and moved?

We will discuss this from a business process perspective in Part II of this book, but firstly we would like to provide the reader with some case studies of organizations that have made substantial progress towards being highly successful using a business process approach.

Chapter 2

Case studies

Introduction

We have included six detailed case studies to provide context for the Management by Process framework. It is all very well reading a book that provides information on what an organization should be doing with regard to the management of its business processes, but we felt it added a significantly deeper aspect to provide case studies on what several successful organizations have actually done.

While you will find a couple of the case studies long, we felt it was better to provide too much detail than not enough. If this is the case, you may wish to skip or skim through them and revisit them after you have completed the book.

It is worth saying that you will need to 'study' or analyse the case studies to determine what worked for the case study organization and if it will work for your organization.

We would like to thank Manual Loos and Saju Madhavan of Citibank Germany, Tony Gardiner of Nedbank, Aveant Home Care and George Diehl of Air Products and Chemical, Inc. for their time in writing or reviewing the case studies.

Case study: Citibank Germany – the 'industrialization' of the consumer division

Most readers would be familiar with the developments in the manufacturing sector over the last 30 or more years. Manufacturing has spent considerable time and effort in the improvement of their production line processes to make them continually more efficient and effective. They have consolidated many manufacturing plants into one or a few, and implemented continuous process improvement programmes, often to the level where it has become a significant part of the culture of the organization. Toyota is the classic case study of the quest for continuous improvement – they call it Kaizan.

The non-manufacturing sector, financial services (banks and insurance organizations), service utilities, government departments and instrumentalities have been much slower to adopt the approach and culture of the manufacturing sector and hence have not achieved the process improvement gains that the manufacturing sector has made.

In 2002 the consumer division of Citibank Germany embarked upon the 'industrialization' of their organization. They adopted the term 'industrialization' to align it with the thinking of the manufacturing sector. This is the story of the gains they have achieved in this part of their organization and it makes compelling reading.

Background

As many readers will know, Germany is a heavily unionized country, with the unions being extremely powerful and having a Worker's Council that works alongside the organizations Board of Directors. All decisions impacting workers needs to be approved by the Worker's Council before they can be implemented.

The Bank is divided into two operating divisions: Consumer and Non-Consumer (investment and corporate). In the consumer division they have 300 branches spread throughout Germany who behaved much the same as most branches throughout the world's Banks, they interacted with customers face-to-face, received transactions via the counter and mail from customers. These transactions were largely administered within the branch or sent to one of the 11 processing centres spread throughout Germany for execution.

Each year the Bank's consumer division receives 11 million documents, makes 160 million payments, 35 million home banking transactions and receives 14 million telephone enquiries.

Business challenges

The Bank faced a number of organizational challenges which included:

1. the need to significantly decrease their expense ratio (expense to revenue) as the initial levels were above the industry average and, in order to effectively compete, this needed to reduce to below the industry average to gain a competitive edge;
2. the need to increase the time spent by branch staff with customers rather than on administrative activities (providing the opportunity of increasing customers service, satisfaction and revenue);
3. the desire to build long and deep relationships with customers;
4. better use of staff while increasing employee satisfaction and providing customers with stable service, even in peak periods;
5. a need to compete in the marketplace and enter new markets.

Like the rest of the German banking community, they were also facing a number of significant market- and economic-based challenges and these included:

1 the demographics of the country, like most western economies, was substantially changing to a significantly older population;
2 this aging population meant a significant shift to a welfare society and the resulting alteration of banking habits;
3 the impact of the country's labor laws (unions) was placing restrictions on the business that were proving difficult;
4 banking products were becoming a commodity;
5 the ethnic population was also changing the way banking was being conducted.

Opportunities

The issue was not only to overcome these challenges, but also to turn them into business opportunities. For example:

1 the aging population provided an opportunity for pension products;
2 the labor laws provided an opportunity to create a relationship with the unions and make them partners in the business;
3 the commoditization of products could create an opportunity of *truly* creating commodity products, for example, personal loans application to approval was reduced from 2 days to 20 minutes;
4 the ethnic population created an opportunity to enable this part of the society to *send money 'home' quickly and easily.*

Results to date

In the first 3.5 years of the Bank's *industrialization* programme the results have been nothing short of spectacular. The consumer division of the Bank has:

1 cut their expense ratio in half; while
2 increasing the customer facing time within the 300 branches to more than 70%. Note: a reputable consultancy completed an industry (German banks) review that determined that branch staff spent 19% of their time with customers. With the removal of the majority of the back-office processing, Citibank's branch staff are now spending more than 70% of their time with customers;
3 errors reduced significantly, from 25–30% to 3–5%, which is considered an acceptable level;
4 staff and customer satisfaction has increased significantly, as measured by surveys.

No matter how you wish to measure its achievement, the Bank has shown the way forward to other banks and industries and created what Michael Porter referred to as a *competitive advantage*.

Approach

It was with this background and attitude that the Chief Operating Officer (COO), and member of the Board, commenced the industrialization of their business. The COO always worked with the tenet, 'perfection is the enemy of good'. In other words, you do not need to obtain perfection in order to be successful. In fact, often 20% effort will often yield 80% of the benefit.

The Banks industrialization process was seen, and understood, to be a *journey* that would take several years to achieve and would need to be addressed on a number of simultaneous fronts. These 'fronts' were considered to be both *internal* and *external*. The *internal* front needed to address internal productivity. In order to create significant productivity gains, both the capacity of the organization and the skill base of the staff needed to be addressed. On the *external* front, the organization needed to address product features, channel features and behaviors.

In this case study we will only be observing the *internal* productivity approach and achievements.

The first step was to address the multiple back-office processing centres and over the period from 2002 to 2003 the Bank consolidated the existing three processing centres into the one service centre. If you are wondering, there are sophisticated business continuity and disaster recovery plans in place within the group.

It should be noted that the Central Service Centre (central processing back-office) is a separate legal entity, and the Bank has outsourced its back-office processing to this separate organization.

The Central Service Centre now comprises 2,000 staff members, approximately 900 of whom are on the telephones (inward and outward bound call centres, collections and telemarketing).

There is a large component of part-time staff and a goal to significantly increase the current number to allow greater flexibility in workforce planning. Germany has maternity and paternity laws that allow for the mother or father to have three years off work after having a child. The Bank has found that many of the mothers wish to return to the workforce much earlier than the three years, but are understandably restricted because of the child. So the Bank has provided the facility for staff to work at hours that suit the organization, the mother and the child.

One of the key goals was to relieve the branch staff from having to complete administrative activities. A few examples of these activities are

- if a person attends a branch to inform them of a customer's death, they are given a special business card and requested to get in touch with the Central Service Centre and the matter is handled in a caring and sensitive manner by appropriately skilled staff;
- credit card limit adjustments;
- all contracts are scanned and centrally stored in an optical format.

The Bank has also negotiated with the Worker's Council and included them in the 'industrialization' process to provide competitive salaries and working conditions. The Worker's Council agreed that the organization could monitor and measure the performance of individuals and provide individual incentives to staff. As at mid-2006 this was at 75% rollout and continuing and will be completed within a year.

This was considered a significant breakthrough with the Worker's Council and the Bank are the only organization that they know of within Germany that has achieved this. More details of this measurement and rewards will be described later in this case study.

The Central Service Centre is also seen as the 'centre for innovation' for the business. It is where they develop and test new products, services and management paradigms.

They have created a department known as the *Engineering and Capacity Management* (ECM) group. The COO's first appointment to this department was a manager who proceeded to implement the 'industrialization' process.

One of the first things the ECM manager and his team created was a *closed-loop industrialization model*. This comprised three components, which are still used today. These three components are

1 Process and Organizational Development
2 Resource Planning and Performance
3 Command Centre.

The department's organization structure and responsibilities are shown in Figure 2.1.

Figure 2.1 shows that the Process and Organizational Development division was responsible for the creation and maintenance of the Bank's business

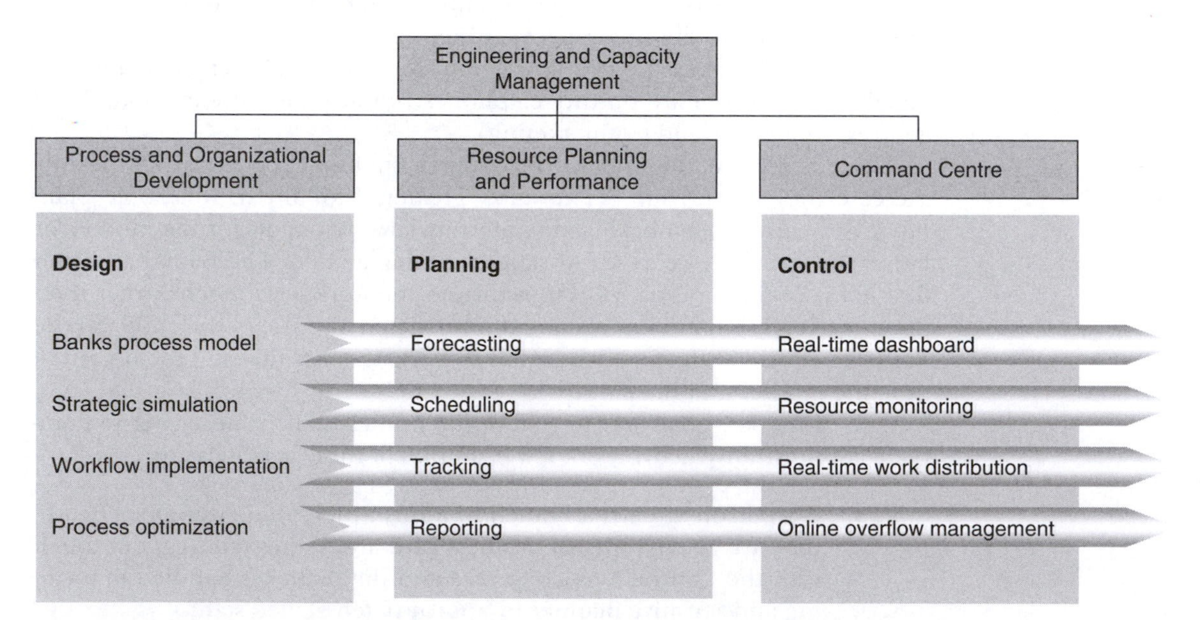

Figure 2.1
Engineering & Capacity Management structure.

process models, the business process simulation function, the implementation of workflow and the continual optimization and improvement of the business processes. Most organization's think that once they have achieved this they have 'done a great job' and only need to create or provide a continuous business process improvement programme or culture.

The Bank realized that this is only the first step and that these activities need to be operationalized within the business if the full realization of operational cost savings is to be achieved, together with an increase in staff and customer satisfaction. This is where the other two divisions come into play.

The Planning division takes the information collected from the workflow system and marries it with the considerable knowledge that they have of their processes (cycle times, transaction volumes and so forth) and then projects staff utilization into the future. In fact this division forecasts business operational staff, which then is provided to operational management to action.

The Command Centre division then provides the real-time feedback and support to operational management and staff. It monitors the inbound and outbound call centres by exception; the workflow system allocates work based upon transaction volumes (real and predicted), staff skill and performance levels and cross-skill factors. This will be discussed in more detail a little later.

Prior to starting their process journey, the ECM department created a vision for how this would be implemented. This resulted in the adoption of the *Industrialization toolset* shown in Figure 2.2.

We will use this model as a means of tracking the evolution of the various phases in the Bank's journey towards the delivery of the desired increase in internal productivity. We will not describe this figure in any detail here as it will be described during the explanation of each phase.

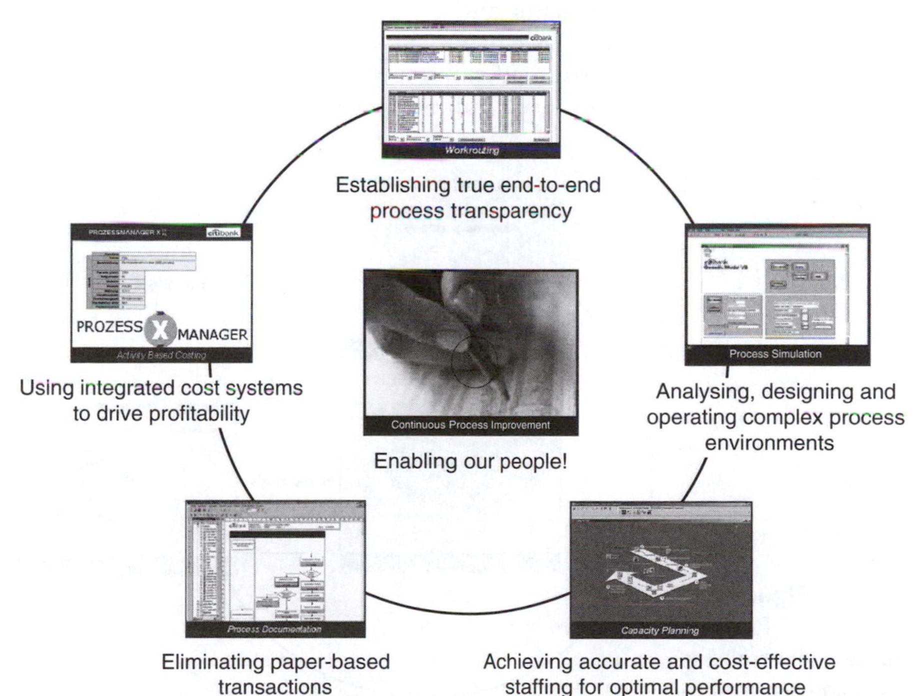

Figure 2.2 Industrialization toolset.

Phase 1: Starting out

How did the consumer division of the Bank start this significant productivity improvement programme of work?

Figure 2.3 shows that the first activities started to include the commencement of the documentation of the current business processes which could then be used for process optimization improvements and simulated to determine both the accuracy of the documented processes and the validation of suggested process improvements prior to their implementation.

Figure 2.4 shows the timelines for these two activities.

Process management training

The first activities started in Q2 (May) 2003 was the creation of *process awareness* which commenced the process of 'enabling our people'.

The Bank would be the first to say that they did not get it perfect first time, but they were smart enough to learn and change the approach as required. In the first instance they provided training to show staff that they were part of a business process, but the day-to-day pressure of business as usual activities meant nothing changed nor happened.

So they changed tact and started to deliver half-day *process awareness* training across the organization. This provided staff with an understanding of the power that the improvement of their business processes would bring to the organization and decreased any fear factors that may be present. They also used specific Bank examples of process improvement throughout the training. But management understood that this alone would not create the changes within the organization that they required. So they created the opportunity, within the half-day process awareness training, for staff to attend a five-day advanced training course. They were only looking for passionate (enthusiastic and ambitious) people to attend the advanced training, so they created specific targeted incentives.

Firstly, staff can only be promoted within the Bank if they have completed the advanced training course.

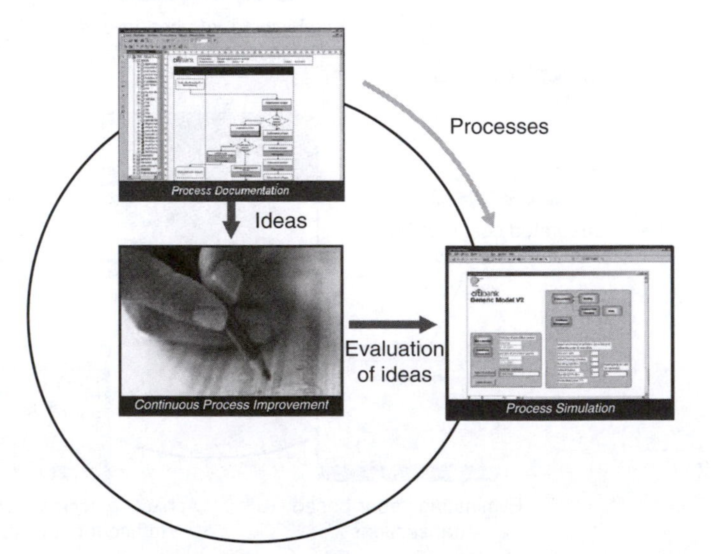

Figure 2.3 Process management training and simulation.

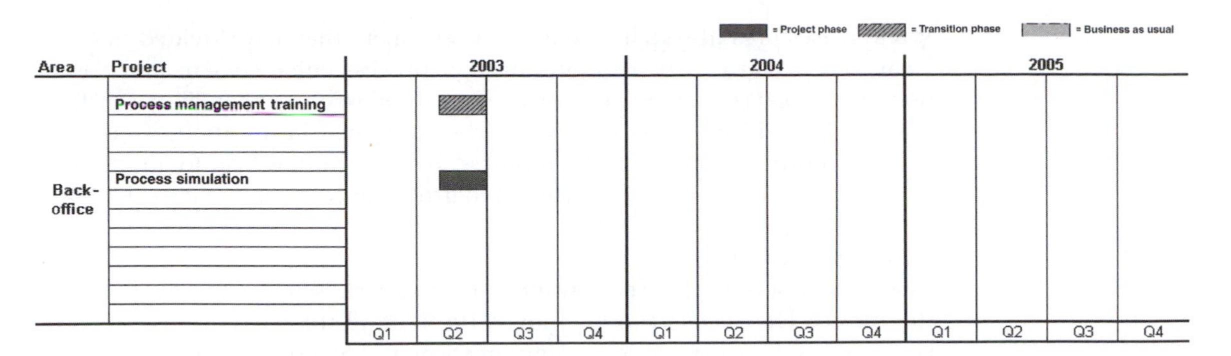

Figure 2.4
Phase 1: Evolution and status of industrialization in Central Service Centre.

Secondly, a staff member could only 'apply' to attend the course and then needed to be accepted. The requirement for acceptance was that the applicant was required to think up a 'breakthrough' project and write a short (one or two page) business case and have the process owner and the manager of the Process and Organization Development division sign it off. To achieve sign-off the business case needed to meet certain criteria which included:

- the project must be capable of being delivered in less than 100 working days
- there must be more than 0.5 full-time equivalents (FTE) savings and this must be sustainable.

The goal of this approach was to create urgency within the business, and obviously to get quick wins. It also provided funding, via the savings, for future business process improvement activities.

Once accepted, there were no more than 8–9 attendees on a course, with two trainers. During the course, the attendees further developed their project plan and became the project manager for their 'breakthrough' project which commenced after the completion of the course. These projects were continuously tracked and the project managers were coached and supported to success during the 100 days of the project. After the successful completion of the course and the project, the success was celebrated and a presentation provided by a Board member.

This approach has resulted in more than 85 certified process experts within the organization and in excess of 100 'breakthrough' projects initiated. These process experts were kept in the business after the completion of the projects to 'spread the word'. If necessary, they could be seconded to other projects in the future.

There were two other significant actions from a project and operational perspective once a project was completed:

1. the savings signed-off in the project business case were withdrawn from future operational budgets;
2. the delivery of the savings (benefits realization) were very well controlled and monitored.

While these 'breakthrough' projects started small, they have yielded benefits (mainly in terms of free capacity through process enhancement and consequent workload reduction and cost avoidance) in excess of €5 million over four years and provided the 'cash cow' for the continued justification of business process improvement within the business. In fact, this has lead to an increase in the budget for the ECM department to grow and continue their work.

Process documentation

The modeling of the current business processes was seen as an investment in the future. They elected to document 100% of all the current business processes and sub-processes, in a common format in a central repository, across the business. They had 150 people across the business involved for 14 months in this exercise. It is important to also understand that they were not involved full-time on this documentation process. This activity was viewed as a pure investment for the future. It could, however, be justified by the quick wins provided by the 'breakthrough' projects, many of which were identified during the course of process documentation.

What business benefits did the Bank gain from documenting their current business processes? They would say that it provided them with:

- a transparency and an end-to-end view of the business processes;
- the ability to provide training, reference material and improved staff induction because of the agreed common format (part of the process architecture);
- more clarity around IT development activities, especially for the implementation of the workrouting system, which came later;
- ability to identify and implement the concept of business process ownership;
- a clear understanding of business process responsibilities;
- an ability to identify process optimization opportunities.

During the process documentation or modeling activities, they linked process steps to corporate policies. This provided the business with a clear understanding of the impact of policy changes and an ability to react quickly.

As part of identifying process optimization opportunities, 'breakthrough projects' were initiated that resulted in quick wins; thereby justifying the investment.

Process simulation

The Bank wished not only to model the current processes, but also to validate the documented outcomes via running process simulations – 'does it (the process) make sense?' This provided 'evidence' that what had been documented was accurate and provided a level of analysis in a complex business process environment.

Process simulation was used as a strategic planning tool to capture the processing activities of a department; using key parameters, such as, average handling time, volumes, employee skills and availability. With the help of simulation, a department head was able to evaluate the different options (without needing to implement them individually) to arrive at the optimal solution for a given situation.

In short, process simulation provided a means to quantify the impact of the proposed redesign changes prior to their implementation within the business.

Process simulation was also be used at a tactical level. By running simulation periodically, using forecasted volumes and average handling times; thereby looking for early warnings, such as, backlogs, periods of under utilization, etc.

Phase 2: Capacity planning

Phase 2 continued with Phase 1 activities and commenced the development and understanding of capacity planning (Figure 2.5).

The time frame for Phase 2 is shown in Figure 2.6.

It is important to understand what is meant by 'capacity planning' in this context. The Bank refer to capacity planning as projecting the future operational business capacity to meet service levels based upon expected future transaction volumes and mix.

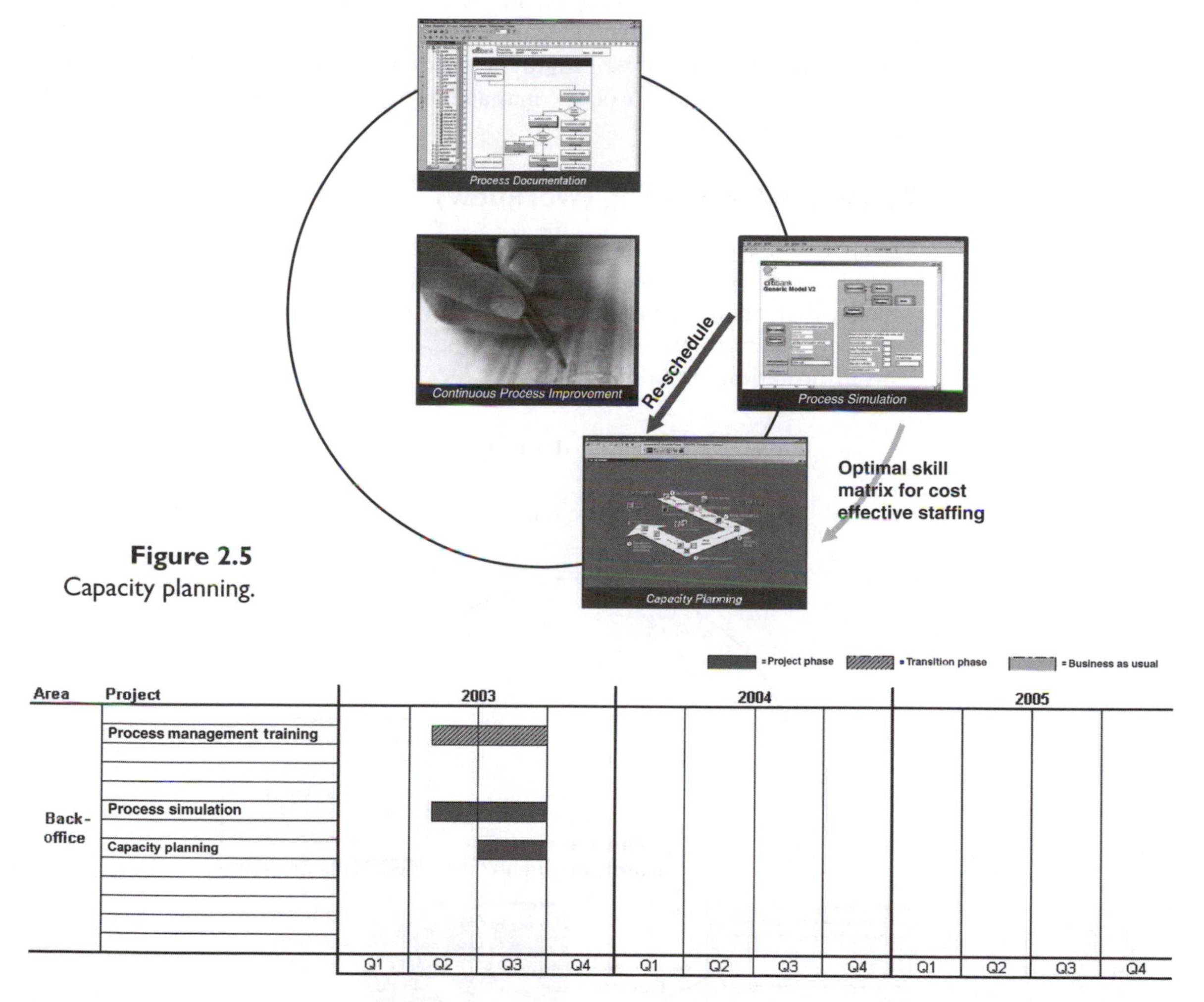

Figure 2.5
Capacity planning.

Figure 2.6
Evolution and status of industrialization in Central Service Centre.

This phase is where the Resource Planning and Performance division of the ECM department project future staff requirements. This is compared to planned staffing levels and provided to line managers, with commentary and recommendations, to 'manage' their part of the business. The inputs into this capacity planning predictions include:

- the expected number of transactions by type
- the number of employees available
- service level targets data, this is, desired handling time
- service level weakness points
- backlogs (in terms of both the number of FTE's and days effort)
- average handling time, by department or team
- staff experience, quality of their work, performance levels and cross-sk illing factors.

As a result of this capacity planning activity, the staffing level of the capacity planning section has decreased from 30 staff members to 4 staff as at January 2005.

Capacity planning is so highly regarded by the Bank that the ECM manager stated that 'process management is nothing without capacity management'.

Phase 3: Workrouting (workflow)

This phase is where the Bank commenced the development and implementation of a workrouting (workflow) solution (Figure 2.7).

Data that provides input onto this environment includes the information gathered for both the capacity planning and process simulation activities. This information included:

- the skill matrix that defines the optimal routing of work packets to employees
- the availability of the employees – in terms of a Control.

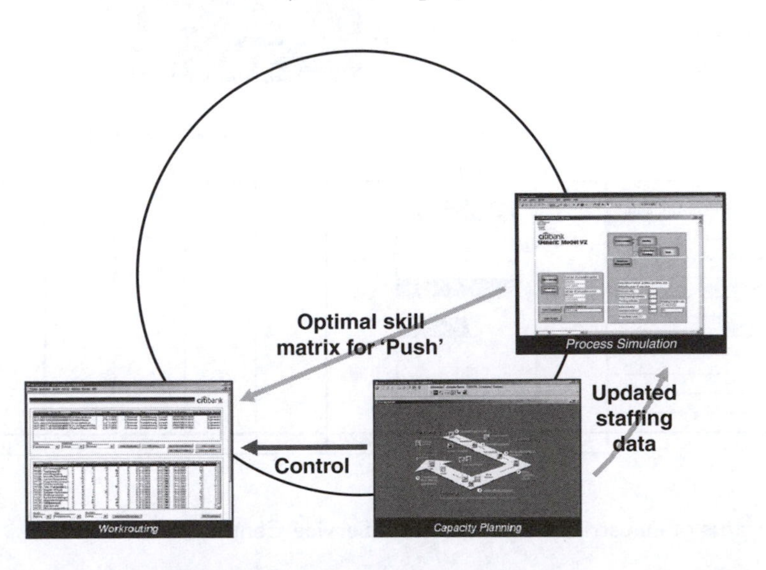

Figure 2.7 Workrouting.

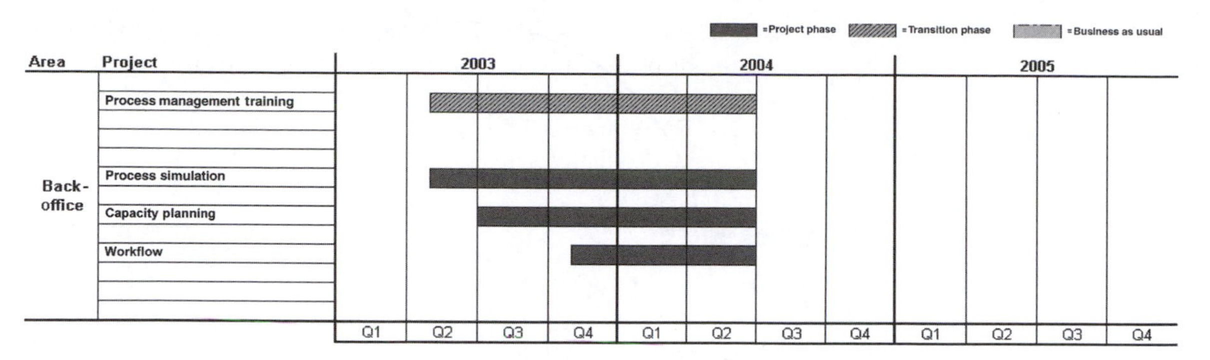

Figure 2.8
Phase 3: Evolution and status of industrialization in Central Service Centre.

Figure 2.8 shows the timeline for this phase.

When implementing workflow, simplistically, there are two traditional methods for allocating work to staff – the 'pull' method or the 'push' method. The 'pull' method allows staff to 'pull' down the number and type of work items that they choose to execute; whereas, the 'push' method 'pushes' individual work items automatically to staff based upon an established set of criteria. In the Bank's experience, the 'push' method has yielded a considerably higher productivity gain compared to the 'pull' method.

The set of criteria established by the Bank for the allocation of work items to staff included:

- processing priority of the transaction type,
- staff skill and experience level, which has an impact on, and an indicator of, the average handling time,
- staff performance level, that is, the quality of their work,
- cross-skilling factors.

Measures were developed and placed in a matrix for staff groups and transaction types. Based on discussions with the team leaders and department heads, transaction types were mapped to staff groups, incorporating a 'priority factor' (priority with which the group of staff would work on that transaction type).

This matrix, together with the current and expected transaction volumes and backlogs determine the allocation of work items to staff members.

The 11 million documents that the Bank receive every year are also imaged, optically stored and attached to work items as they 'move' around the organization.

Phase 4: The Bank's process model

Phase 4 developed the ability for gathering all the data from the previous activities collected into a data warehouse for use and feedback into the rest of the 'toolset'. Accurate processing data (times, backlogs, delays, staff performance levels and so forth) allow for more accurate and improved processes, together with an ability to continually improve business processes (Figure 2.9).

The period from mid-2004 to the end of Q1 2005 saw the completion of the end-to-end process modeling across the organization (Figure 2.10).

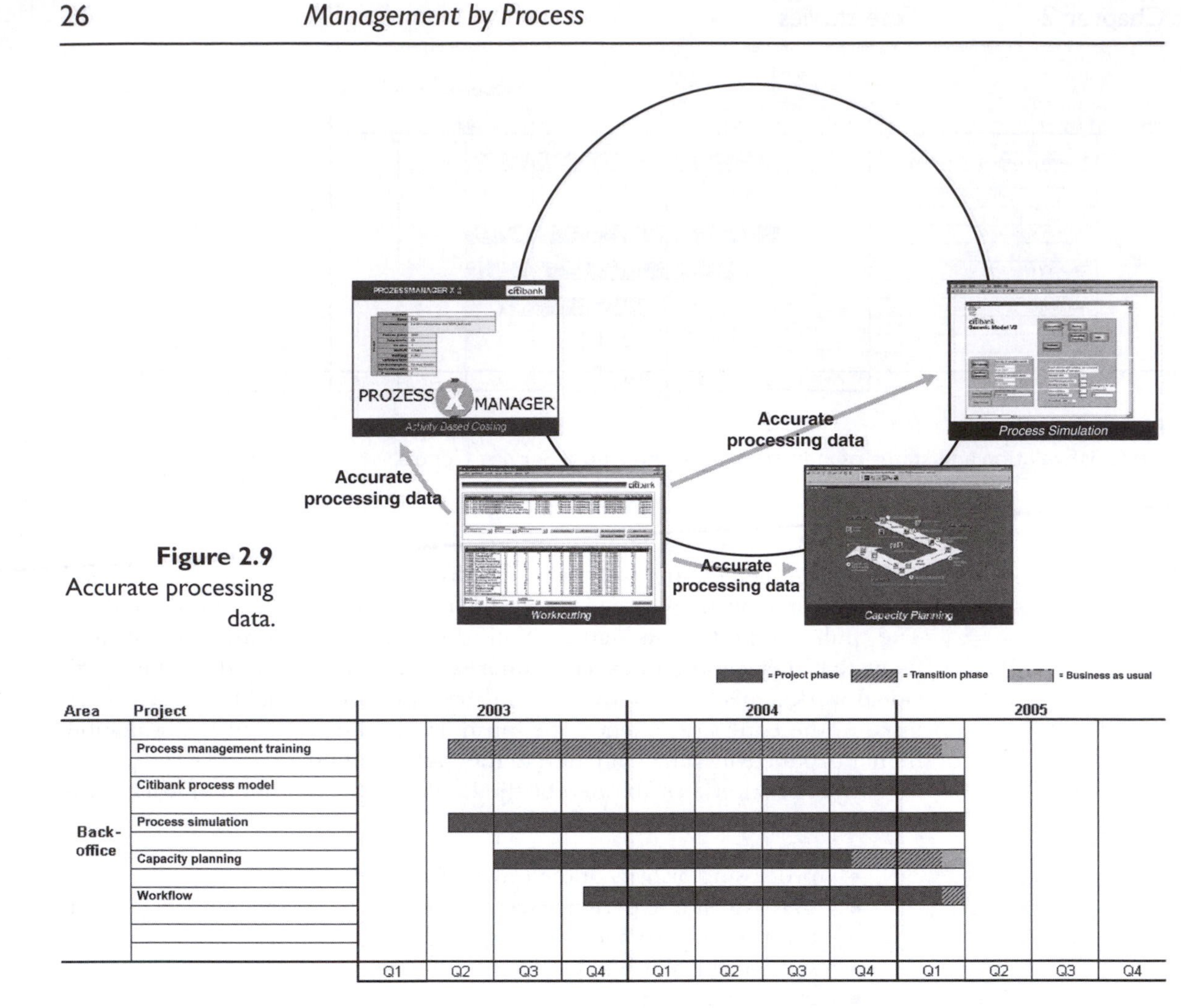

Figure 2.9
Accurate processing data.

Figure 2.10
Phase 4: Evolution and status of industrialization in Central Service Centre.

This phase delivered:

- a transparency of the current end-to-end processes
- a structured, consistent and complete process landscape
- process documentation for the back-office was completed on an activity level.

As this information further evolves, the plan is to make it available in an integrated central web-based process portal (database) with version control and audit trails, along the lines shown in Figure 2.11.

As can be seen in Figure 2.11 the intranet portal includes:

- the organization chart
- all documented business processes; these will be split by operational, control and business processes
- products and services offered by the business

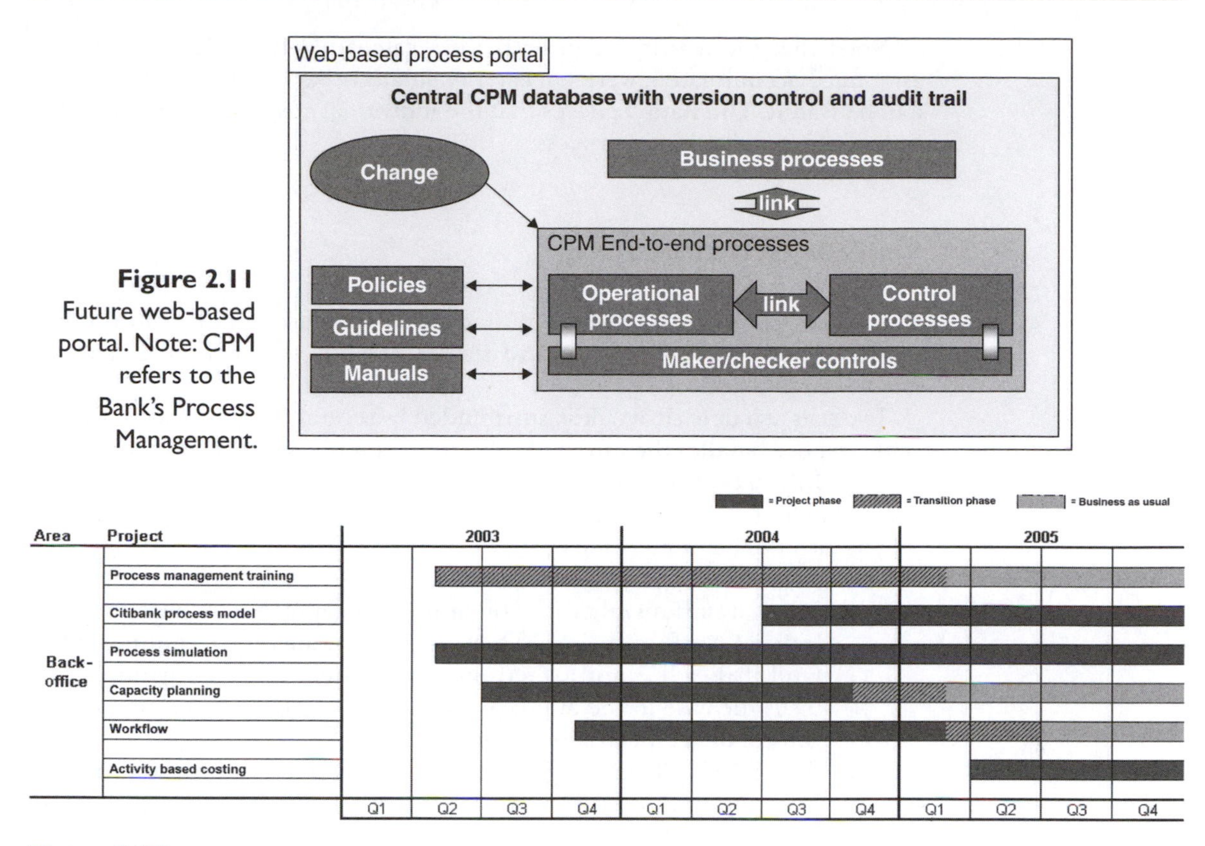

Figure 2.11
Future web-based portal. Note: CPM refers to the Bank's Process Management.

Figure 2.12
Phase 5: Evolution and status of industrialization in Central Service Centre.

- policies
- procedures
- guidelines.

Phase 5: Activity-based costing

The remained of 2005 (Q2 to Q4) comprised the completion of activity-based costing on the business processes to:

- determine the true end-to-end process, product and channel costs
- build the basis for future financial benchmarking

with the intention to extend the Activity-Based Management for profitability analysis (Figure 2.12).

Process activities were grouped, or boxed, into those activities that were to be measured. The Bank has been measuring many of the activities for some years now and was now able to match appropriate process models to the existing historical measures.

The costs of non-process activities (meetings, etc) were distributed across all processes to provide a 'true' cost. These non-process activity costs were regarded as costs over which they had no control. If a process is optimized and the cost changes, then the non-process costs were redistributed.

Note: that the activity-based costs were calculated from the 'bottom up' and, once accumulated, were checked (validated) back to the actual costs of a department. The Bank believes that the individual costing provides a reflection of the 'true' cost of a process.

On-going monitoring

The on-going monitoring of the business processes are handled, in the first instance, by the Command Centre staff. Currently this station is manned from 8 am to 9 pm.

Two staff at a time sit at a desk surrounded by several computer screens. One staff member handles the inbound call centre and the other the outbound call centre. Only exceptions are displayed on the screens, where an exception is defined as an event outside the established goals or targets. Examples of these targets include the monitoring of:

- when a staff member logs on at the call centre is compared to their official work start time. There is an allowance of 8 minutes grace and if they are not logged on by this time, the Command Centre calls the case to the attention of the SPOC (single point of contact) within the respective department/area;
- length of call time. Again, if the call length is outside the normal call duration, a team lead is called to go and provide assistance to the staff member;
- call wrap-up time. This is handled in a similar manner to the above call length situation.

While this appears to be a little like 'big brother', the aim is to assist staff and managers in the execution of their tasks and the management of their business and it is conducted and received in a positive manner.

Individual staff achievements of targets are reflected in the bonuses they receive. These results are published only to the individual, his/her team leader and department head. At first the bonus system had a floor amount, which would be given to all staff, and the team leader and/or department head would decide the additional bonus on top. With the performance-based bonus system, it could so happen that some staff receives close to nothing whereas the high achievers almost double their salaries.

The targets include: quality, achieving service levels, selling products, selling appointments to a branch, did the customer actually attend at the branch, what did they buy and so forth. Staff *and branches* are also provided with a disincentive in case of bad appointment quality or high cancellation rates. The reporting on these targets is fully automated.

Authors comments

One of the authors spent some time with the 'industrialization' team in Germany and they were an impressive and intelligent team of people. The reason for the visit was to learn from their success and compare the organizations approach to the project implementation framework outlined in previous book (Jeston and Nelis, 2008).

The authors would like to reflect on the organizations obvious success and highlight a number of activities and approaches that we believe ensured the successful outcomes.

1. The initial process improvement team was small and built on their success.
2. The business process improvement programme had the support of the senior executive of the organization.
3. Process awareness workshops were held across the organization to 'spread the word' of the benefits of improvement. When the initial training model was not as successful as expected, the team quickly adapted and created the short half-day session.
4. Providing an opportunity beyond these short sessions to a longer training course allowed ambitious and enthusiastic staff to contribute. Making applicants for these longer training courses apply (via a business case) by identifying a 'real' business process improvement opportunity within the business meant that only committed staff applied.
5. Furthermore, coupling the process management training to the career path provided the right impetus among the staff of the organization.
6. The type of projects (100 days, no IT changes) meant that quick wins were implemented, thus justifying continued process improvement activities.
7. We like the fact that the applicant's managers signed-off on the business case. This meant that only genuine business cases were approved and that the manager actually had to realize the business benefits identified in the business case.

These activities meant that the business process improvement programme was commenced with a high probability of success and with significant commitment and support from the organizations managers and staff.

The project team, staff and entire organization should be very proud of themselves and their achievements.

Case study: Wealth management organization

As you will find when you read this case study, it is really two case studies in one. The first shows the traditional business process improvement project (we have called it the review project) which yields significant potential gains to the organization.

The second was a pilot where the consultant was asked to work with staff over a very short period of time (three weeks) to address a specific issue for the organization. This was also viewed as a pilot for the review project. This specific business issue was costing the organization a significant level of business revenue; the time frame allowed was very short to launch the pilot; and the outcome was amazing, increasing customer satisfaction *and* reducing operational costs – all this with no changes to IT applications or business processes – there was simply no time. It is a true demonstration of the power of people.

Background

This was a small business unit of a much larger bank, insurance and wealth management organization. The business unit had in excess of 100 FTE staff and an annual operating budget of $10 million. The business unit develops financial products and sells them via a network of independent intermediaries (financial advisers, banks, building societies and so forth). It develops and markets for the superannuation or pension fund marketplace.

In December 2003 two business units were joined together following an acquisition.

Historically, the main business unit was always client focused. Staff were able to process any request from members or advisers within the one geographic region. The newly acquired business unit was functionally based – that is, a staff member would only process one type of transaction but would do so for every client and adviser regardless of their location.

These two business units were left that way under the one management structure until August 2004, when it was decided to merge them under the one client-focused structure. It was fully understood by management that there would be a significant training exercise required to achieve this new client-focused approach. So customer-focused training was delivered to provide staff with the necessary skills. The business was structured on a geographical regional basis and all work coming into the organization was segregated by region and then processed. Staff had the dual responsibility for processing the transaction and liaising with customers. Management believed that the additional costs associated with a non-production line approach would be outweighed by the benefits of increased job satisfaction for staff, greater accountability and better customer satisfaction.

Business challenge

One of the prime goals of this new approach was to improve customer service within six months which was not achieved, despite extensive on-the-job and formal customer service training. This was further compounded by an unacceptably high expense ratio, work duplication, increasing error rates and service standards not being achieved.

The fact that the two original business units were still being administered on different system application platforms (legacy applications) did not help the situation. It required everyone to be fully familiar with both systems. So a decision had been made to rationalize onto one platform but that was still likely to be several years away.

The customer service in the original business unit, which was originally rated as very good, began to fade. The customer service in the newly acquired business unit, which was rated below average before the amalgamation, failed to improve.

Review project

Approach

The overall business unit decided to approach three service providers in the area of business process management to provide proposals of how to remedy the situation. They did not include the incumbent software workflow vendor in the short list as previous experience indicated that the vendor concentrated solely on the workflow software whereas the business unit wanted to review the processes from an end-to-end perspective and then review and understand the business implications.

The management was fully committed to the success of this review project and demonstrated this by the level of management time allocated to the project. The business unit manager and his three Client Service Managers allocated two days a week to the project for the project duration. Other staff including, team leaders and senior fund administrators were also involved in workshops, reviewing outcomes and revising estimates – a commitment of six staff from the overall business unit of 100+ for about 20% of their time over the project.

The selected BPM consultancy recommended a phased approach with several 'gates' to allow the business to stop at any stage, ensure it was receiving value for money and that the project was delivering as expected. There were four phases recommended, with the expected duration indicated below:

Approach (phases):

1. Discovery – 2 weeks
2. Understand (As Is) – <5 weeks
3. Innovate (To Be) – ~6 weeks
4. Final Report – 2 weeks

These phases were delivered by the BPM consultancy with a lead consultant and one senior consultant for the duration of the engagement. The consultants:

- meet with key stakeholders
- conducted workshops (process execution staff and management)
- ensured that all necessary stakeholders who were external to the main business unit were fully engaged. This included the finance department and IT
- process modeled the current processes and the proposed new processes
- completed significant metrics analysis.

Table 2.1 provides a more detailed explanation of each phase.

Table 2.1
Detailed phase steps

Phase	Steps or activities	Deliverables
Phase 1: Discovery	This phase had an elapsed time of two weeks and comprised the following steps: 1 high level 'walk through' with processing staff 2 brief discussion with key members of the management team 3 development and agreement of the High Level Value Chain 4 development and agreement of the Process Selection Matrix, including a list of identified processes 5 high level gathering of associated metrics 6 define and agree Project Scope 7 production of a detailed project plan for Phase 2 8 production of a draft plan for Phase 3	Deliverables included: 1 High Level Value Chain 2 Process Selection Matrix, including a list of processes and high level metrics 3 a detailed project plan for Phase 2 4 draft plan for Phase 3 5 a report on Phase 1
Phase 2: Understand	This phase had an elapsed time of five weeks and comprised the following activities: 1 process modeled the current processes at a level that enabled the Innovate phase to be completed 2 gathered baseline metrics from which to measure improvement activities 3 completed an appropriate level of Root Cause analysis to ensure an understanding of the base cause of an issue and not just 'treating the symptoms' 4 completed a list of process and business stakeholders and engaged with them 5 completed a list of major process issues, as determined by the business 6 identified Innovate phase priorities 7 identification of opportunities for quick wins 8 validate and agree on the implementation of quick wins	Deliverables included: 1 process models of the current state of the processes 2 appropriate metrics sufficient to establish a baseline for future process improvement measurement 3 a report on the Phase 2

(Continued)

Table 2.1 (*Continued*)

Phase	Steps or activities	Deliverables
Phase 3: Innovate	This phase had an elapsed time of six weeks and comprised the following activities: 1 a Management Workshop to agree the goals of the processes and scope of the Innovate Phase 2 process model the new redesigned processes at a level to enable their implementation and/or automation 3 complete an appropriate level of metrics (process cost) analysis to demonstrate an indication of the likely process improvement and potential cost savings 4 identification of quick wins that can be implemented in a short time frame 5 validate the feasibility of the proposed redesigned process options 6 identify the benefits of these redesign options and update or create a business plan if required 7 prepare a presentation for senior executives 8 gain approvals for the redesigned processes	Deliverables included: 1 list of agreed process goals 2 process models of the redesign processes 3 key findings and an analysis of the process human touchpoints 4 list of recommendations 5 recommended Phase 4 project structure and roles and responsibilities 6 a preliminary risk analysis 7 suggested next steps 8 list of agreed Critical Success Factors 9 presentation to senior executives 10 phase report
Phase 4: Final report	Write the final client report for the project	Final report

At the conclusion of each phase the client received a report that provided an opportunity to evaluate the phase and stop or redirect the project if necessary. The time frames indicated were met and there was no project overrun from a time or budgetary perspective.

The small project team engaged the staff and management significantly in the project. An early engagement was considered an essential part of the people change management aspects of the project and necessary to achieve the results the business needed.

Project findings

The Discovery and Understand phases revealed a number of significant pieces of information:

- there were 12 quick win opportunities identified and the business commenced implementing 6 of these immediately

- there were 20 primary business processes, which represented 95+% of the business operational costs, the other processes were considered to be of no great consequence in the context of this project
- the top 4 processes accounted for 65% of the operational costs
- a further 23% of costs were spent on enquiries and complaints
- checking of the transactions processed was considered necessary because of the exceedingly high error rates and this checking activities accounted for 17.5% of staff time at a cost of $455,000 per annum.

This obviously provided an indication of where to expend effort in the Innovate phase.

After the Discovery and Understand phases were complete it was necessary to agree with business management what approach they wished to take to the redesign of the business processes in the Innovate phase. It was considered as essential to 'set the rules' as to the options or scenarios they wished to consider during the process redesign activities. The guidelines needed to cover both the type of options that were available and the associated timeframes. The following was agreed:

- outsourcing was to be considered
- the timeframes were to be 6 months and 18 months
- both a 'portfolio' or 'process' based transaction processing approach was to be considered, as long as the process goals were achieved
- the project was not to be bound by the current culture of the organization
- the project was not to be constrained by the existing staff roles and organization structure
- technology to be considered could include, and not be limited by: web-based transactions or portals, intelligent forms software, optical character readers, imaging solutions, legacy system enhancements (although there was a two-year lag from the vendor on such changes) and a 'full' implementation of Business Process Management System (BPMS)
- needed to consider legislation, security, compliance and fraud
- quality needed to be build *into* the processes
- simplicity wherever possible.

With these guidelines in mind, the business decided upon three scenarios for the Innovate phase:

1. Scenario 1: had a timeframe of 18 months delivery and was to include a complete redesign and redevelop the workflow application; including the introduction of image processing (document scanning and optical storage); and a 'full' BPMS solution, which would provide: a business rules engine, business activity monitoring (BAM) and flexible reporting. To the organization, a full BPMS solution would provide business agility, process control and management.
2. Scenario 2: had a timeframe of 18 months delivery and was to include a complete redesign and redevelopment of the workflow

application; no imaging; and hard-coded reporting and process management (BAM).

3 Scenario 3: had a timeframe of 6 months delivery and was to include the redesign of the existing processes to accommodate the existing workflow solution and with no changes allowed to the legacy application.

The Innovate phase comprised three primary activities:

- the creation of redesigned process models (for each scenario) which were predominantly produced in workshops
- the agreement of a new set of process metrics and business operational costs (these were devised by business staff and only facilitated by the consultants)
- the agreement of new staff roles and organization structure, via the creation of a new people capability matrix (refer to Jeston and Nelis, 2008, p. 136 and 161).

These activities, when compared to the baseline metrics calculated in the Understand phase, showed the results in Table 2.2. Note that detailed costings were only completed for Scenarios 1 and 2.

In the calculation of the process metrics staff were divided into productive and non-productive.

The 'raw' numbers for the savings were even better than indicated in Table 2.2, but they made no allowance for contingency, such as, peaks in transaction volumes which occur in this industry; sick leave, annual leave and so forth. So the project team consulted with the business to allow management to determine the level of contingency they needed in the business. The metrics were then recast accordingly and the outcomes shown in Table 2.2.

As can be seen, Scenario 1 provided the business with the opportunity of a 39% saving in their annual operating budget, while Scenario 2 provided a 29% opportunity.

Table 2.2
Cost comparisons

Scenario	Contingency (%)	Budget ($)	Number of FTE's			Utilization (%)	% Decrease to operating budget
			Productive	Non-productive	Total		
Now	0	10.0 million	85.5	24	108.5	100	
1	33	6.1 million	49	12	61 (49%)	83	39
2	34	7.1 million	62	12	74 (45%)	88	29

Figure 2.13 shows graphically what the business unit needed to be able to achieve from a metrics perspective in order to mount a business case to management and is a simple representation of Table 2.2.

Benefit Gap A measures the gains associated with the Scenario 3 and the quick wins from the Understand phase. The business did not spend any project time on estimating these. They simply made sense, did not cost much to implement, so they just implemented them.

Benefit Gap B shows the savings, $2.9 million, from the implementation of the Scenario 2.

Benefit Gap C shows the additional benefits, $1.0 million, to be gained from the implementation of Scenario 1.

Benefit Gap D is obviously the total of both Benefit Gap B and C, which totals $3.9 million.

How was this achieved?

The business and the consultants followed the appropriate parts of the 7FE Project Framework as outlined in our previous book (Jeston and Nelis, 2008).

While we have discussed several of the activities completed in achieving these savings, there are a number we should highlight:

Process metrics analysis

The metrics analysis was considered to be a critical part of the project. Data was gathered from budgets, organization charts, actual transaction volumes and actual and estimated process times. This data was gathered during workshops, discussions with management, observation and various SQL queries against the application systems.

It was considered important that the management and staff 'own' the estimates of the time to process the various business transactions. These estimates were gathered in workshops, validated with 30% of the staff who did not attend the workshops, and then extrapolated and reconciled against the actual staff employed in the business to ensure there accuracy.

People capability analysis matrix

This matrix is completed during the Understand phase, to reflect the current situation, and then again in the Innovate phase to take account of the

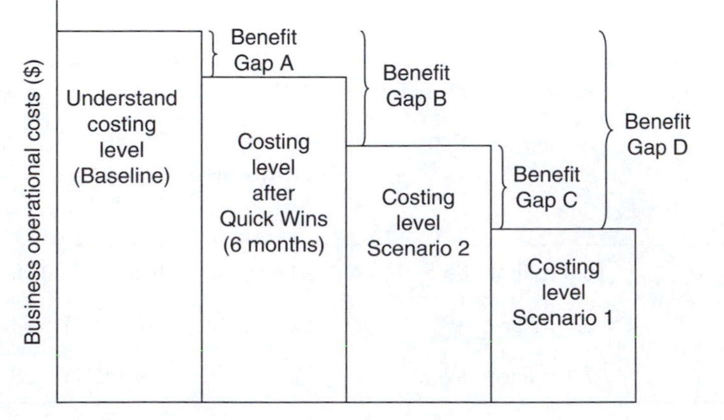

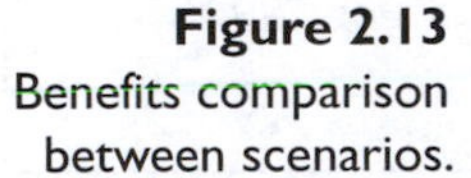
Figure 2.13
Benefits comparison between scenarios.

Key processes \ Knowledge capabilities/ skills required	Ability to sell to customers	Communication skills	Data entry skills	Dealing with difficult customers
Notification	2	2	3	1
Assessment	1	1	3	1
Approval	3	2	3	1
Payment	2	2	3	2
Finalization	2	3	1	1

Figure 2.14 People capability matrix.

new redesigned processes. A comparison between the two can yield interesting information and any gaps need to be well documented and understood because it could make a difference in the creation of new roles and responsibilities within the business, as well as implementation training for staff.

Figure 2.14 provides an example of how this matrix could be completed. The horizontal axis represents the core skills or competencies required by each of the processes to complete the tasks or activities. The vertical axis represents the end-to-end process model, group of processes or individual processes. These core competencies are then rated on the simple basis of 1, 2 or 3, where 1 is a mandatory core competency and 3 is desirable but not essential (Jeston and Nelis, 2008).

In the case of this organization, there was a substantial difference between the current and future state requirements for the staff competencies. The information provided at the commencement of this case study in the background section indicated that the organization expected staff to be both relationship managers with specific customers and have detailed administrative processing skills. This was like wanting staff to both a 'salesperson' and an 'accountant' at the same time. While some people do possess both these skills simultaneously, they are a rare breed!

We will discuss how the roles were changed and implemented in the next section when we discuss the pilot implementation.

Organization structure

The organization structure was reviewed as a result of the obvious savings in operational costs. The business did not need as many managers and team leaders with the reduction in staff levels. The current structure is shown in Figure 2.15.

It can be seen that there was a National Manager, four senior managers and ten team leaders for the 100+ staff. The project team projected that only 61 staff were required. This was discussed and with the business National Manager and it was agreed that the new proposed structure would look like Figure 2.16.

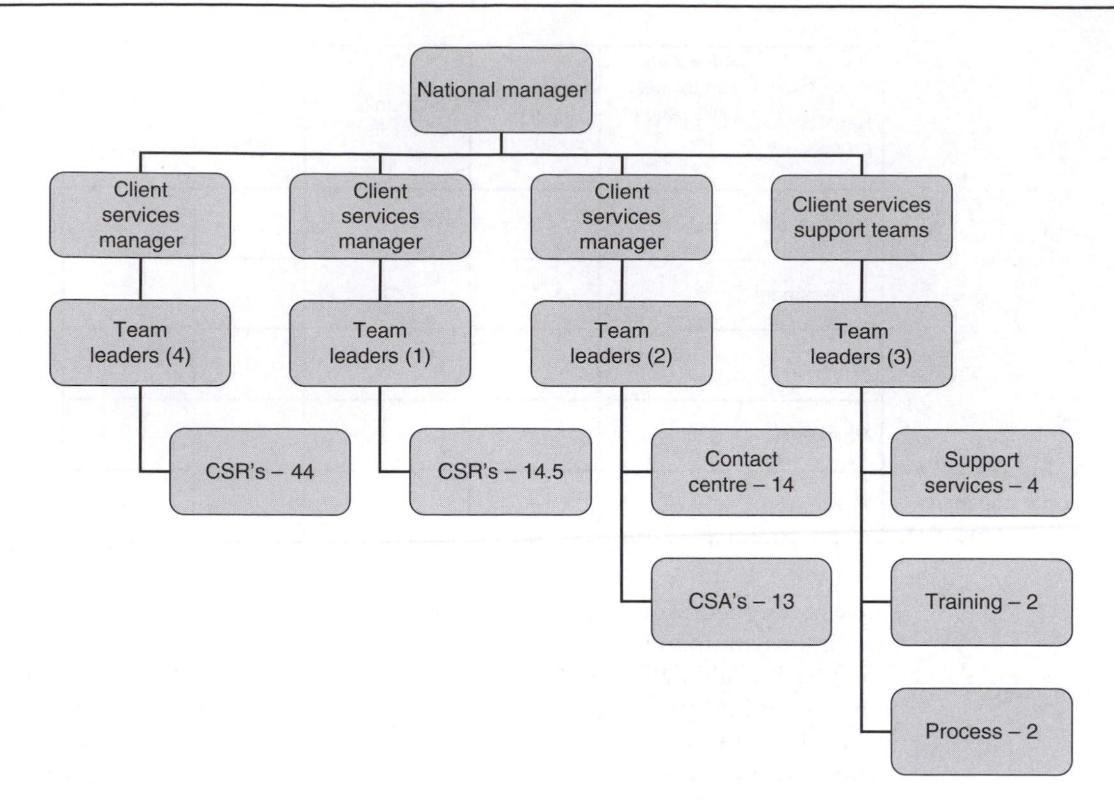

Figure 2.15
Current organization structure.

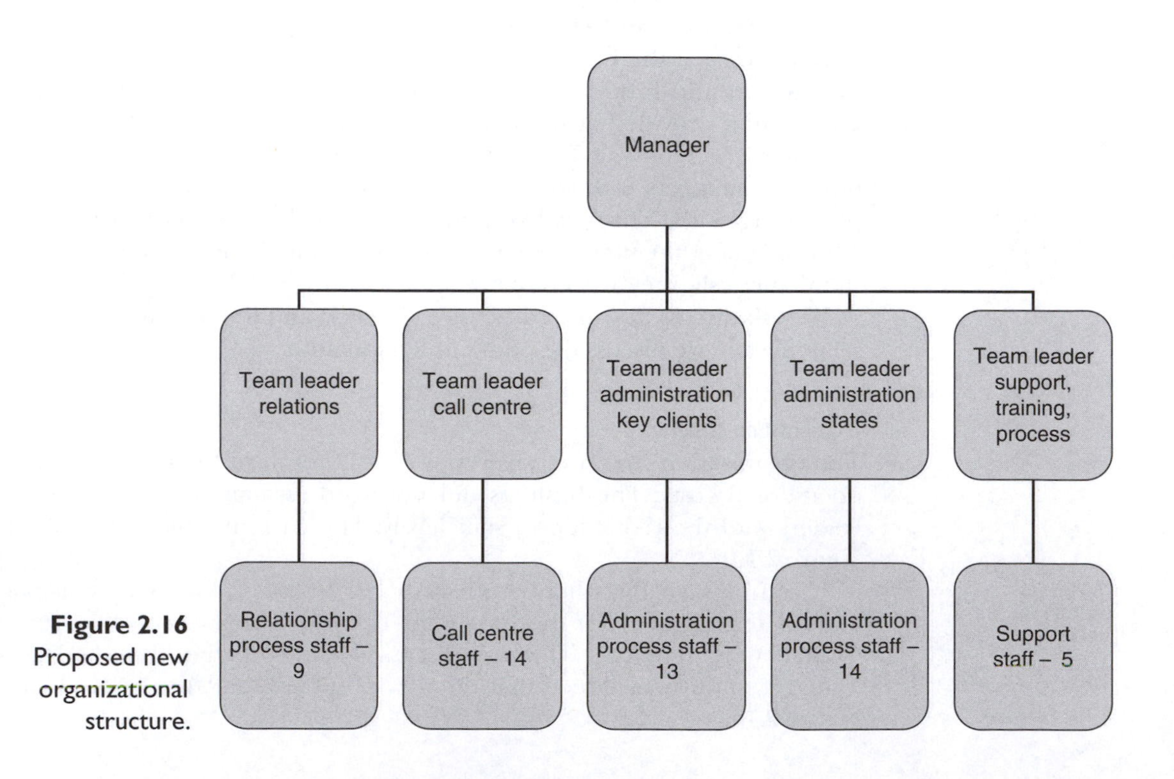

Figure 2.16
Proposed new organizational structure.

Figure 2.16 not only shows the reduction from five to one senior manager, but also the team leaders reducing from ten to five. It also shows the change in roles as a result of the People Capability Matrix analysis. The roles were split into a Relationship role and an Administrative role, thus enabling staff to specialize and to be accountable for specific activities. This was particularly important in the pilot that we will review next.

Business pilot: Starting small

About 12 months before the business improvement project mentioned above commenced, the business created and launched a new financial product that was distributed via intermediaries (financial advisors). The product was considered by all to be the best on the market. Management decided that there would be minimal system or process changes until the success of the launch was known. This resulted in a significant increase in the number of manual process work-arounds to accommodate the new product.

Impact on the business

The work-arounds, coupled with the existing processing problems of the business unit, had a compounding affect.

While the selling intermediaries acknowledge the product was the best on the market and initially supported it, they withdrew their support as the administrative work-arounds and considerable processing delays and errors materialized, which resulted in a significant reduction in revenue.

Administration was very poor with significant backlogs and agreed service level agreements not being met (two to three month processing times when it should have been a matter of a few days). This approach nearly brought down the success of the new product.

Organizational challenges

As all this was occurring, the senior management of the organization almost entirely changed. A new COO was employed and he completed a trip around the country to meet the distributors/financial advisors and listen to them; and they were extremely critical of the administration of this product. This also coincided with the release of the business process improvement project report: immediate action was required.

The COO decided to implement part of the recommendations of the project review specifically for this product; this involved establishing a separate team to administer the product, and one of the authors was requested to project manage and implement the new team.

Pilot brief

The pilot project brief proposed establishing, within three weeks, a new specialist team, going live four days before the end of the financial year. (Superannuation or pension plan transactions *significantly* increase for several weeks after the end of the financial year-end.)

The *only* constraint was that there would be no time to implement any changes to the business processes or the IT systems.

Approach

On 5 June the pilot to establish a stand-alone team was commenced and went live on 26 June, just before the end of the financial year (30 June). Staff were specifically recruited by invitation only and then were interviewed. There was some initial pushback by staff to the change but eventually several accepted the new roles, and this had a snowballing effect as the project became successful.

The suggestions of the business improvement project review – such as changes in the definitions of roles and the metrics knowledge gained – were implemented, with the one existing role split into a 'relationship role' and a 'technical or administrative' role.

Results

Results were almost immediate: Improved customer service and a substantial reduction in error rates despite still relying on the existing processes, IT applications systems and manual work-arounds.

The team not only successfully took over all outstanding transactions (backlog) just prior to 30 June (year end), they also handled the year-end increase in volumes (600–800%), cleared all existing backlogs and increased customer service – *within five weeks* from the start.

The impact to the financial adviser distribution network was significant and immediate. It resulted in an increase in business volumes – in fact, *300–400% within the next four weeks* – which lead to an increase in the number of staff in the pilot team, to cope with the increased workload and revenue.

What was done to achieve this success?

Staff were selected for their attitude more than anything else. They did not know the product, and it was a very complex product. The team leader and one of the authors (consultant) spent time briefing staff around what was expected of them. The three tenets that were non-negotiable were *ownership, accountability, and responsibility (OAR).* The team leader and the consultant ensured that staff had a say in the way the team was constructed and managed – obviously only up to a point. At the end of the day, the team leader was responsible for the performance, and if she failed, they all failed.

The other non-negotiable tenet for the team was *zero external errors.* The error rate up to this point, throughout the business, was so high that the business had introduced double-checking of transactions – and it still resulted in an unacceptable high error rate to customers. The old saying that 'team' stands for Together Everyone Achieves More was very true in this situation. A genuinely empowered 'team' environment was encouraged where each team member knew that to be individually successful was only achievable if they were all successful. So stronger team members assisted weaker members and helped to build their skill levels.

The pilot was also the first time that staff was truly managed on a process performance basis.

In order to measure staff performance, the estimated process execution times from the project review discussed earlier, was used to establish the performance measurement targets for individual team members and the team as a whole.

The outcome was that, with a focus on performance measurement and managing the processes, the team processed the transactions significantly faster than originally thought possible, and certainly faster than the rest of the business. This resulted – with the staff agreement – in the *increasing* of their targets.

Summary

In summary, the results of the pilot were that:

1 within five weeks the team had:

- a learned the product
- b cleared all of the processing backlog
- c coped with the 600–800% increase in transaction volume due to year end

2 and, four weeks later:

- a customer satisfaction and confidence had increased to the extent that the business had increased 300–400%
- b staff were exceeding the processing targets set for them, and they were increased.

All this with a zero external error rate, which was down from the 30% prior to the pilot.

Authors comments

We would like to make comments and observations about both the business improvement project and the pilot:

Business improvement review project

1. Making staff and management available for the Understand and Innovate workshops was initially very difficult when the existing transaction processing backlogs were considered. The senior management made a significant commitment of subject matter experts (SMEs) to allow the project to be successful. This is a typical issue with most business process improvement projects. Solutions are for the project team to be understanding of the impact upon the business operations; conducting workshops at times that are convenient to the business; the business may need to budget for the backfilling of staff with temporary staff to allow the SMEs to be available.
2. The project team should not create the process metrics but simply facilitate the information from the business. The project team should check and validate the metrics to ensure they ‘make sense’. This allows the business to unequivocally ‘own’ the outcomes.
3. When completing process redesign, do not become so focused on the redesign process that the organizational impacts upon staff

roles and responsibilities is not reviewed and evaluated. This activity made a significant difference within this organization.

Pilot project

4 The only *real* activity that took place in the pilot was a sensible allocation of roles and responsibilities; expectation setting and management; treating staff as true business partners and thus empowering them. We refer to this as 'people change management'.
5 While it is interesting how performance changes as soon as it is measured and becomes visible, it can be a two-edged sword, and therefore needs to be thoroughly thought through before it is implemented, if you are to achieve your desired outcomes.
6 The pilot could only have been as successful as it was because of the valuable information available from the business review process improvement project, which provided the knowledge of the need to change the roles; metrics analysis and had provided the confidence in the team to complete the pilot.

Case study: Nedbank South Africa

Driving towards a process culture in the Technology and Operations division of Nedbank South Africa.

(Dr. Tony Gardiner)

Organization background

Nedbank Group Limited operates as one of the four largest banking groups in South Africa. The group offers a wide range of wholesale and retail banking services through three main business clusters: Nedbank Corporate; Nedbank Capital; and Nedbank Retail – the principal services include corporate and retail banking, property finance, investment banking, private banking, foreign exchange and securities trading.

The group also generates income from private equity, credit card acquiring and processing services, custodial services, collective investments, trust administration, asset management services and bancassurance.

Nedbank Groups head office is in Johannesburg, with large operational centres in Cape Town and Durban and an extensive branch and support network throughout southern Africa.

Some key 2006 statistics include:

- Total Assets – R 424,912 million (approximately USD 60,702 million)
- Total number of employees – 24,034

- Cost to income ratio – 58.2%
- Market share:
  - Total assets 19.8%
  - Total deposits 21.0%
  - Mortgage lending 20.2%
  - Credit card 12.0%
  - Branches 1310
  - Self-service terminals 1599

Overview

In the mid-1990s, faced with all the opportunity of new and available technologies, yet constrained with the challenges of Y2K, large legacy systems and the distributed operational environment of the past, a large part of the Bank embarked on a massive drive in search of business process excellence.

Many of the details of this journey are probably similar, if not the same, as in many other organizations with regard to process improvement. What was unique was the emphasis placed on the cultural transformation of the entire organization, the drive to become a 'process-centric' organization, and ultimately the emergence of the Business Process Maturity Model (BPMM), recently adopted as an international standard by the Object Management Group.

What is described in this case study is the organizational change journey, the events that led up to the development of the BPMM, our experiences with the model and the original Capability Maturity Models (CMM), where it took the organization and where the organization is now. The content of the model will not be dealt with here, but hopefully the context that we provide will give you some understanding of the full intent and use of the model, and its value in the realization of a common 'process culture'.

During the latter part of the journey, the organization experienced some major difficulties, not as a result of the process journey, but rather as the cumulative result of negative issues which converged at a crucial time (macro-economic, acquisition and strategic). This resulted in changes in leadership, structure and focus with consequent effects on the process journey.

Case study background

Around 1995 the operations and technology of almost the entire bank were joined together into one 'mega-division' which was referred to as the Technology and Operations division (Figure 2.17).

The technology half had grown from the single central IT function which developed, maintained and operated the old legacy mainframe systems and was responsible for almost all the IT systems in the bank.

The operations half had quite different origins – from de-centralized beginnings it had moved to four big regional hubs which were responsible for much of the back-office processing functions that had once been part of the branch network. These included:

- the Financial Control Teams that controlled all the branch suspense accounts;

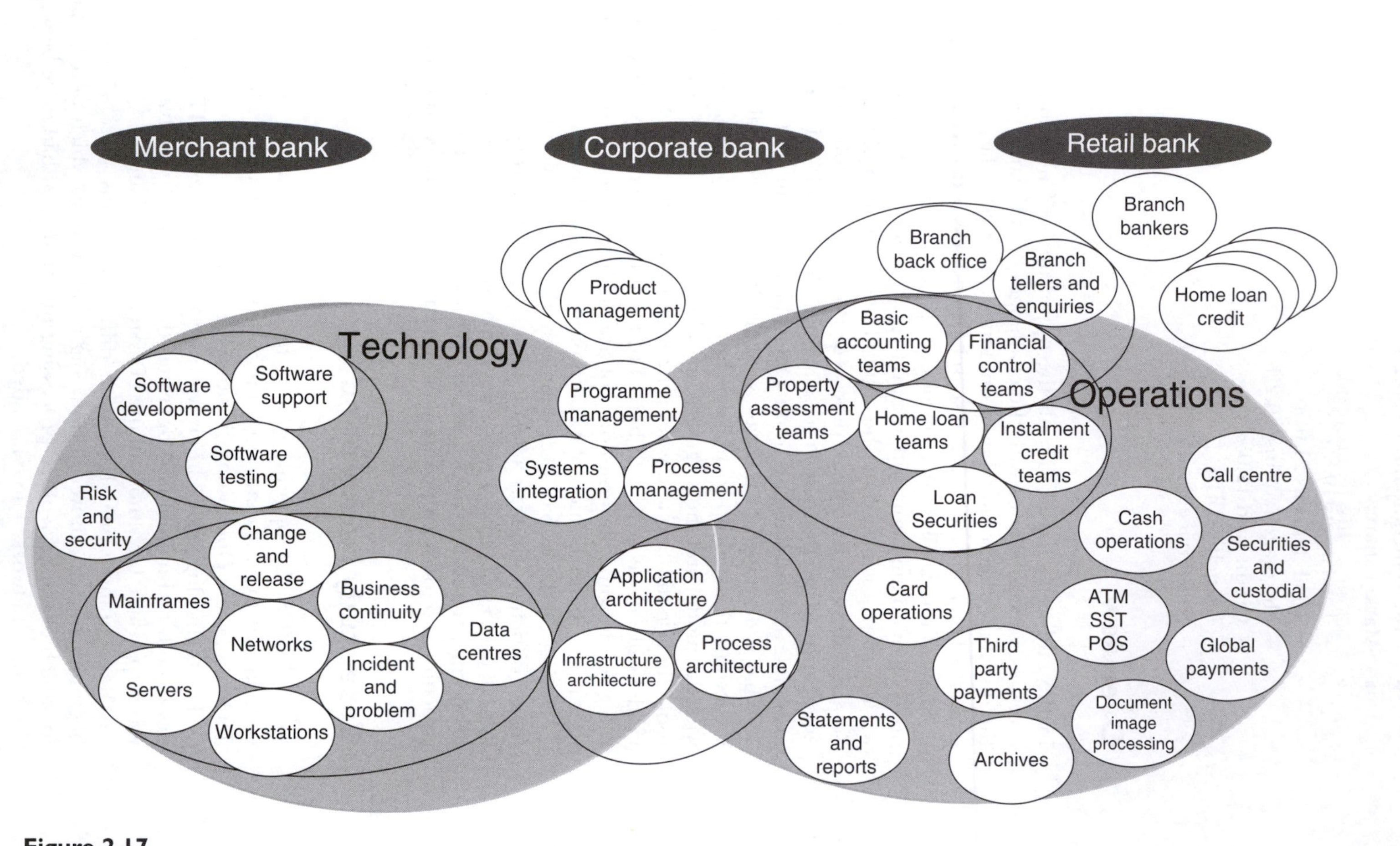

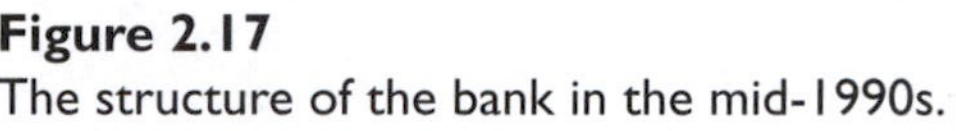
Figure 2.17
The structure of the bank in the mid-1990s.

- the Basic Accounting Teams that reconciled all the un-postable transaction entries;
- as well as the teams responsible for processing mortgage loan and vehicle finance applications.

As image processing technology was implemented the function was also centralized alongside the hubs and the many distributed paper processing centres (PODs) diminished.

The other functions were all largely centralized, often with smaller satellite functions in the other regions associated in one or other way with the big regional hubs.

Supporting 'business functions', like credit, were also undergoing centralization at the time, both geographically and across products.

Branch back-office centralization was taken a step further than normal when the branches were split into separate sales and servicing functions – the remaining back-office functions, the enquiries and the teller functions reported in to the Operations hubs and the selling or relationship management function reported in to the Business.

Although much of the centralization had been enabled by the continuing automation of different functions and activities, the creation of a single Technology and Operations division primed the organization to capitalize on optimizing all the many different operational processes and procedures.

For this reason the Process Management function was created and grew into a considerable size over the next number of years.

In the Innovation space – those responsible for leading, managing and facilitating technology projects – Programme Management was responsible for the project management, Systems Integration for facilitating the technical project requirements; and between Process and a function known as Product Management the business cases, business requirements and all the supporting processes and documentation were developed.

Business challenges

Given the disparate and decentralized beginnings many problems existed in each of the different areas.

Operations – most of the common problems were typical of an operationally inefficient or process immature organization, namely:

- many different ways of doing things; differences between and within regions; generally no documentation of any of the processes or procedures, either locally or end-to end; processes relied on pockets of knowledge, and local experts;
- little if any measurement of the processing time, quality, capacity and no understanding of the levels of error and rework, and capacity management was through experience rather than planning;
- hand-offs were a huge cause of frustration, with little quality checking between functions, little common understanding of the requirements of different functions and little understanding of the dependencies of different functions;
- no planning, no prioritization processes.

Technology – all the same problems as in Operations, and in addition the demands of Innovation, combined with little reinvestment in the assets.

Innovation – the problems included: huge demand; the constraints of the looming Y2K; little project control, large scope creep and never enough resources. With the advent of Y2K (constraints) and the many new and now proven technologies (demand), the bank was faced with multiple conversion, integration, enhancement and new technology projects. Many, if not most of these projects, were running over budget, not delivering the expected benefits, the victims of continual scope creep, errors and the resultant rework at all stages and often well into the Software Development Life Cycle. With the many concurrent projects, the organization was continually short on resources and facilities (particularly with integrated user acceptance testing). There were thus major bottlenecks throughout the project lifecycle and a continual struggle with prioritization. There was little understanding of the possible synergy between the different projects or project streams and that contributed to much duplication of effort and the resource conflict already mentioned; and although the projects and initiatives were not failing, they were costing the organization much more and getting much less done than should have been the case.

Opportunities

In short, non-standard and non-aligned processes, both in the Innovation life cycle and in the Operational areas, were recognized as one of the critical reasons why these problems were being experienced:

- to apply the technology – we needed to understand, to align and to standardize the business processes.
- to deliver the projects – whether automating, integrating or converting:
  - we required a common and collective understanding of what happens in the different parts of the business (an end-to-end view)
  - we needed to document the business processes; and to specify the rules, activities, dependencies, inputs and outputs of all the multiple back-office functions
- similarly to improve performance in the operational areas – as a result of the increased focus on improved customer service, lower costs, improved turn-around times, better quality and reduced risk – standard and aligned processes were required.

Results to date

- consecutive years of zero percentage cost increases in the operational areas – large impact on the cost to income ratio;
- large savings achieved in the process re-engineering projects;
- the development of an integrated process architecture containing all the process assets;

- the organizational understanding and management of time, cost, quality and risk in all operational areas;
- the attainment of CMMI Level 4 maturity in the software development areas with the concomitant benefits in quality and productivity;
- the attainment of CMMI Level 2 maturity in the full Innovation area;
- the development and application of the Business Process Management Maturity.

Approach

The whole Technology and Operations 'mega-division' did not come about by accident; the CIO had fought for the control of that 'turf'. He had a vision and big plans to realize this vision.

The first step towards that vision was the drive to a process organization; and so in 1996 a process management function was created – with, very soon, a few hundred dedicated process engineers.

Establishing the process management function

The evolution of the process management function involved the following.

Firstly, the adoption and implementation of a recognized case tool to define and analyse the processes. What did exist in the way of procedures had been captured in individual diagrams, training manuals and hardcopy documents. Early on the choice was made to use a formal case tool (in this case the ARIS toolset) for the analysis and modeling of all processes.

In order to logically store the many models that were being generated the development of a five-level enterprise architecture for the whole of the bank (roll-up from task level to the top ten functional areas of a financial services organization) was started; it went through many stages eventually evolving into what was referred to as the Integrated Architecture Framework, and included process, product and technology.

To produce re-usable, integrated, high-quality process models required the development of modeling standards. Standards were selected and adopted, filters were set up and semantic checks were run on the models, release control and formal database management were later areas of focus. With that many process engineers released into the line areas, a multiplicity of business process models were generated at Levels 4 and 5 of detail, describing activities and tasks.

Once an initial understanding and documentation of the business processes had taken place and the obvious areas of opportunity had been identified, business cases describing the opportunities for process redesign and automation were developed. The focus was on savings, and harvesting of the 'low-hanging fruit', and the business cases required sign-off from the business line managers, in order to ensure the realization of the benefits.

A huge demand grew for process resources to do all types of things – any problem was viewed as a process problem and anything that needed fixing was viewed as a process. This required processes for how the function worked – how work was accepted, clarified, estimated, planned and scheduled; it also allowed the cost of process improvement to be established.

The re-engineering of processes, the application of new technologies and the integration of new and old systems to do things differently (as well as the automation of processes through the application of work flow tools) were both primary areas of focus and effort. They resulted in the process resources fulfilling the full business analyst role on projects (with the exception of product feature and functionality), developing the business requirements and the process models to support them.

Once the business processes were modeled, the opportunity for reliable measurement became realistic. Time standard calculating software was used to formally calculate and set up time standards for the key functions in the operational processes. The time standards applied to the key function volumes allowed the calculation of activity-based costs and the development of a full activity-based costing model.

Business line management involvement with process improvement initiatives was encouraged through the establishment of Process Enhancement Groups (PEG's). Initially they played a significant role in developing a common understanding and standard view of the operational processes, and thereafter were able to identify the incremental improvements from the floor. In parallel and supportive to the development of the PEG's was a strong effort to educate and inculcate a process culture throughout the operational areas. The effort involved workshops, coaching and training which will be described in more detail later.

The Six Sigma methodology was also introduced. A strong Six Sigma initiative was launched with experts being flown in from the United States and a large number of process engineers being trained up to green belt level and the whole Technology and Operations division being introduced to the approach.

In short, all the right things were being done with a huge level of enthusiasm and energy, but they were difficult to keep aligned and it generated its own set of problems and complexity (Figure 2.18).

Using a case tool to document As-Is processes has challenges – the discipline of formal business process modeling requires a high level of skill, a detailed understanding of the process and produces technical flows which are not easily understood, or easily translatable into a process user friendly format. The process models are also not easily moved between different toolsets, neither for producing simple user friendly flows nor for detailed software design or automation.

Bottom up modeling of processes required the rapid development of a top down view of where to put the different process models. At the time, few process architectures for financial services institutions were readily available for reference, and the development of the banks own process architecture took too long relative to the delivery of the process models. Consequently most process models did not integrate or link easily to provide the end-to-end or higher level views. Many also did not adhere to the quality standards, with different interpretations of the levels of detail and usage of the standards.

The process models also reflected the differences between parts of the bank (different regions), different product variants and reflected different attempts to redesign the processes (multiple versions). With no formal structure or understanding of how to store these models, a massive database of models evolved that could only be navigated by the individual process engineers that had created the particular models in their own working space. It became 'affectionately' known as 'spaghetti' modeling and was an enormous barrier, preventing the development of standard, re-usable process models.

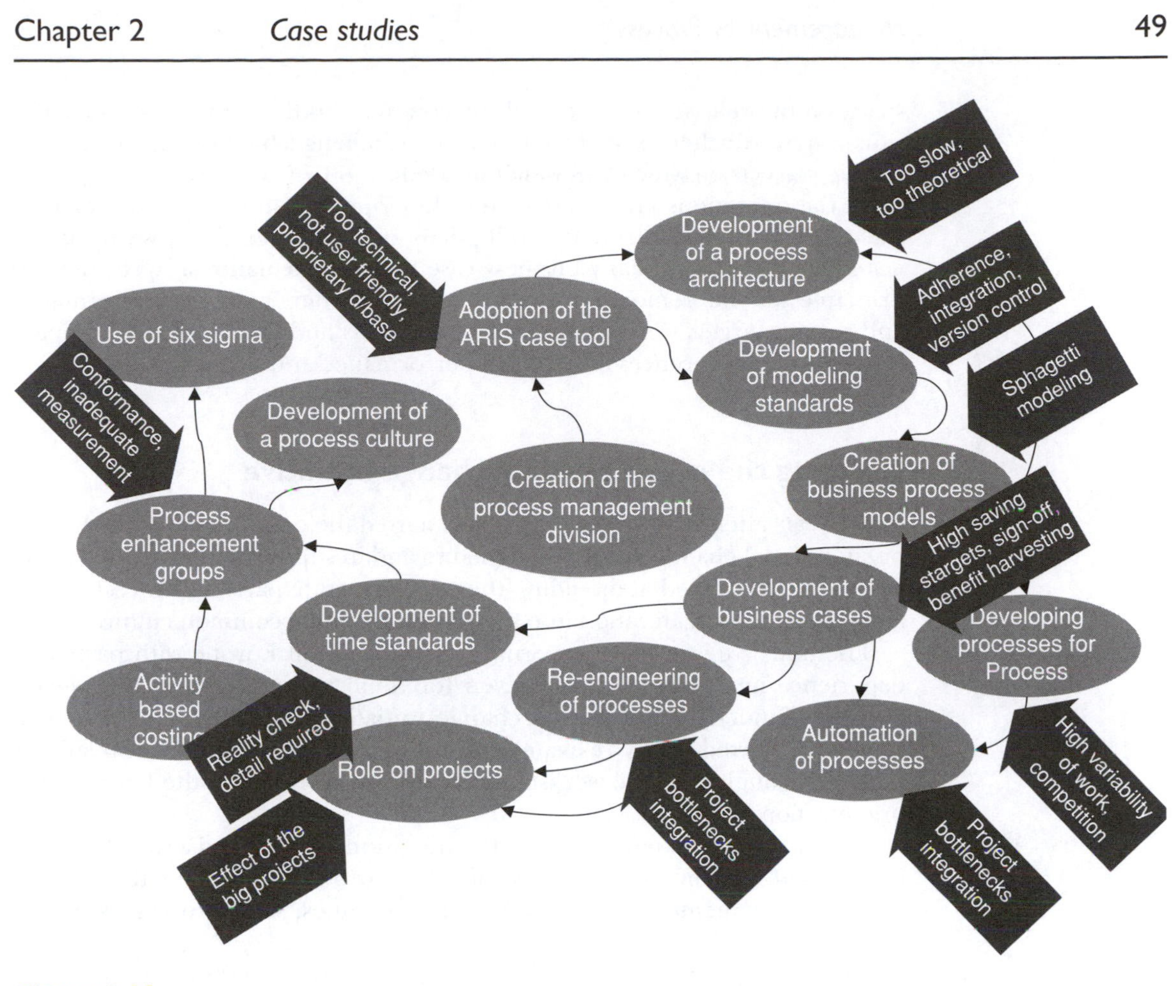

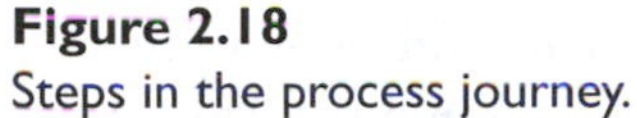
Figure 2.18
Steps in the process journey.

Contributing to the creation of the multiple process model versions was the strong focus on cost recovery. The Process Management division was measured on the business benefit achieved and thus the focus was on the development of business cases with less focus on achieving full process governance.

A healthy tension developed between business line managers (responsible for business case sign-off) and the process engineers (responsible for business case development). Business line managers were held accountable for realizing the savings and the process engineers for finding the savings. Business cases often ended up highly contested rather than co-operative efforts.

The fact that all problems were very soon seen as process problems meant that the process engineers were pulled in many different directions acting as generalist consultants rather than specialists. Although many problems were successfully addressed, this compromised the establishment of process baselines and the training and development of a methodology required to perform high quality process engineering work.

Driving re-engineering and automation solutions meant being part of a larger project team. This in turn meant exposure to all the problems of the Innovation area, namely fighting for priority, for resources, for facilities, for space on the release schedule – all the common bottlenecks of the time. The situation was further exacerbated by the unchallengeable priority of Y2K and the big projects on which Y2K was dependent.

In the operations environment the development and agreement of time standards, complicated by the multiplicity of process versions, were further tested as the basis of many business case savings calculations. Agreement in principle was far removed from commitment when budgets, performance and savings targets were based on the standards and inter-regional comparisons exposed differences in financial performance and savings realization.

Creating the organizational change initiative

Three years after the process journey was started the organization kicked off an organizational change initiative, aligned to and in support of the process goals, but specifically aimed at moulding the once disparate parts of the Technology and Operations organization into a single entity with a common culture.

The initiative was led by external facilitators from Europe with particular experience in this domain, and even for them this was one of the biggest and best funded organizational change initiatives that they had ever been part of. It started off by re-examining the vision, values and critical success factors, through multiple workshops conducted through all the layers of the organization.

It very quickly expanded beyond the vision, values and critical success factors and consumed the process journey to be described in the context of six core elements that included structure, values, staff, process, skills and systems.

For each of these streams senior management working groups were set up, with supporting task forces, to examine the problems and design and propose solutions. The entire senior management structure was divided up into 'gangs' with a rotating overall responsibility for driving the initiative forward. At regular intervals Top 200, Top 500 and Top 1000 staff events were organized in which progress was reported, support was reinvigorated and the organization re-focused and excited by the journey.

Process was a core component and the idea and vision of a process culture was developed and driven out across the organization. Workshops were designed with facilitators and coaches and run across all levels of the organization. The workshops used the concept of building a small wind-up machine made from paper, elastic and a cotton reel – called a Tok Tokkie. From this the teams learnt the basics of good process design and management with undertones of lean and Six Sigma, and had a lot of fun doing it.

Process Management created its own specific vision in support of the Technology and Operations vision – 'to create world class processes and through them to substantially improve productivity'. The process vision was described in a 'COPI' diagram which detailed what the function was in place to achieve and how it achieved it by detailing the Controls, the Inputs (what was needed to do it), the Processes (how to do it) and the Outputs (what was done) required to deliver that vision.

These diagrams were established for all the different areas within the Technology and Operations organization and forced each area to re-examine

Striving for a Process culture

We will know we have it when:

- *Every process is identified, defined and always followed* to ensure our customers receive consistent service – **we do it the same way every time**.
- *Continuous improvement is a way of life* – **there simply must be a better way**.
- *Finding a problem is a happy occurrence* – we don't hide or deny it, it focuses our attention on what we need to do better – **problems present opportunities to improve**.
- *We all understand end-to-end processes* and each of us knows how we add value – **everyone understands the 'big picture'**.
- *Everyone understands their own customers and suppliers* and has agreements on customer needs – **customer pull, not supplier push**.
- *We manage the exceptions, the processes manage themselves*: quality is part of every process – **quality is passed on, not inspected**.
- *When we make a decision, we consider our customers and suppliers*, not only ourselves – **look right, look left**.
- *We use technology to further improve the process*, and not only make processes faster – **engineer, automate and digitize**.
- *Feedback, accurate and appropriate measurement* are catalysts for improvement – **we're passionate about metrics**.
- *We understand our processes*so well, they are second nature – **process is a way of life!**

'There simply must be a better way'
TECHNOLOGY and OPERATIONS DIVISION

Figure 2.19 Vision of a process culture.

what they were doing, with what and for whom. It provided a good preliminary view for establishing what processes were required and in place for each area.

In addition to the specific process vision, a broader organizational process vision was developed and communicated, specifically aimed at supporting the drive to achieving a common process culture. In effect it described what was meant by a process culture and provided the guidance to the behaviors expected in a process culture – this is seen in Figure 2.19.

One of the key drivers for the whole initiative was the visionary objective of becoming a global service provider. The intent was to provide a flexible and scalable processing platform that could form the basis for an outsourcing business. The rationale was the perceived move in the industry from vertical to horizontal integration, the ongoing banking imperative and the resultant quest for quality and the view that global players would have ever increasing scale advantage. The direct advantages that it believed it would bring to the bank were the gains in scale that would reduce the banks costs, the ability to spread the technical investment costs and the creation of substantial new value streams for the Bank. One of the unique underpinning capabilities that the organization believed would allow it to move successfully into this arena was its implementation of a process culture. The implementation of a process culture would enable:

- the detailed understanding of end-to-end process costs;
- improved processes to be implemented faster;
- enhanced customer service, leading to market share gains;
- translocation of operations to other geographies.

The arrival of CMM

This provides the picture of how serious the organization was about business process during this period, however, the problems in achieving the process culture that were described above had remained.

The investment in process had been made, and scores of process engineers had been sent into the operational areas to understand and document the processes. They had achieved results – good business cases, smart solutions, successful technology projects, slicker processes and significant cost savings (quick wins) – but we still had problems. The same problems in Innovation of resourcing, cost, scope and priority remained, if not more so now that there were so many more new solution proposals, and still only product- or channel-based views of the processes.

The spaghetti modeling had resulted in many detailed and rigorously modeled processes that we could not integrate across the different functions. To store all the modeled processes in a process architecture would literally require re-modeling and the dream of standard, aligned, re-usable process components was not being realized. Process engineers were still getting pulled in all directions, and they lacked the discipline of storing, maintaining and re-using the processes.

In effect they were achieving lots of benefit and success (quick wins, and technology solutions), but all focused in specific different areas with nothing to hold it all together. It was, to a large extent silo-based with no overall organizational improvement. The organization lacked a roadmap or framework that could be applied across the entire Technology and Operations division to guide the improvement efforts and achieve the sustainability of the process effort that was required.

During this period the Software Development area had discovered and adopted the CMM for Software. They established their own process improvement initiative, based on the guidance of the framework, and very quickly started to yield results. Most importantly the approach made sense, particularly with respect to the problems being experienced elsewhere in the organization. The model provided a checklist of practices that need to be performed, it provided guidance on the governance and assurance of processes, had a strong focus on measurement (including statistical process control), provided a roadmap to improvement and most importantly, provided the means of reaching a common culture.

This approach produced results, massive improvements in process performance and quality of output in a relatively short period of time. Over a three year period, productivity (function points per staff month) showed an improvement of 28% in the second year, and 46% in the third year. Whilst some of the increase can be attributed to the inaccuracy of data in the early stages of the process journey there is no doubt that large gains were made. In quality, expressed as defects per function point peer review, many defects were picked up, but initially a lot of them were minor or cosmetic in nature (spelling mistakes, grammar and the like). As time went on, however, the quality of reviews improved and although fewer defects were picked up they were more serious in nature. On reflection the severity of the defect should have been reported on. In quality, expressed as defects per function point produced, a truer reflection of the improvement in quality

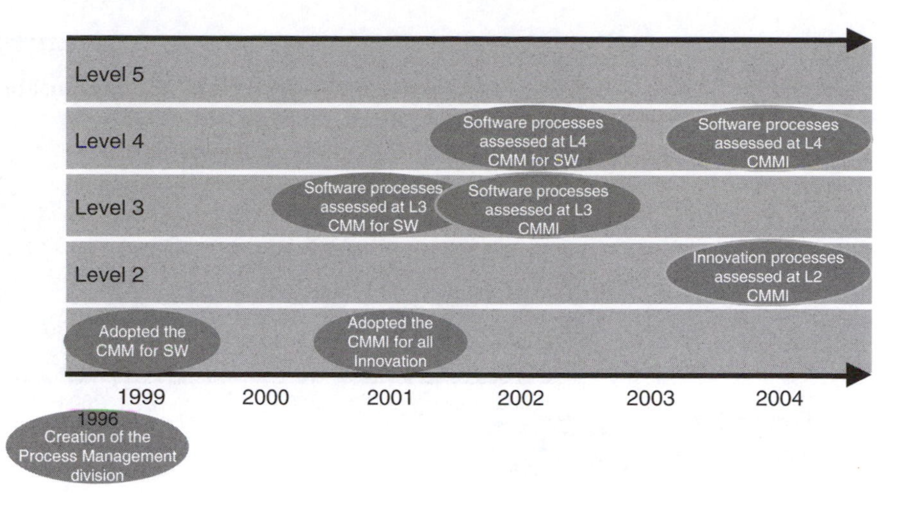

Figure 2.20 Implementation of capability maturity.

was obtained – whereas defects per function point peer review showed a 40% and 96% decrease, this reflected as a 20% and 25% decrease.

The Software Development area adopted the CMM for Software during 1999 and by 2001 had been assessed at a Level 3 Maturity. By 2002 it had been assessed at Level 4 and the improvement described above was measured between 2002 and 2004 while operating as a Level 4 software development organization. Very few organizations in the world were operating at that level of process maturity and capability at that time and the benefits were obvious to all.

With the release of the CMMI, for use across the entire Innovation area and life cycle, the organization was quick to move into a sponsored initiative to achieve the same sorts of benefits on a greater scale. With many more disciplines and areas involved, the initiative was far bigger and the results were not achieved as quickly. The CMMI was adopted by the entire organization in 2001, the Software Development area was assessed at Level 3 in 2002 and Level 4 in 2004, whereas the entire Innovation area was assessed at Level 2 in 2004.

The adoption of CMM for Software and the CMMI is depicted in Figure 2.20.

The formulation of the BPMM

However, the entire Innovation area still only represented 20% of the cost of Technology and Operations, and what was important was how the other 80% could be addressed. The efficacy of the approach had been tried and tested. Millions were being spent on business process management in the operational areas, but no internationally accepted frameworks for launching, managing and evaluating these business process initiatives could be found – because none existed at that time. The organization was participating in the business process outsourcing boom both as a supplier and a customer – but had no means of evaluating the capabilities of itself or others as suppliers. It was faced with many new emerging technologies, but like most other banks did not have the confidence or capability to respond to the early stage market. Excellence in management is rooted in fact based decision-making – modeling and measuring is what matters.

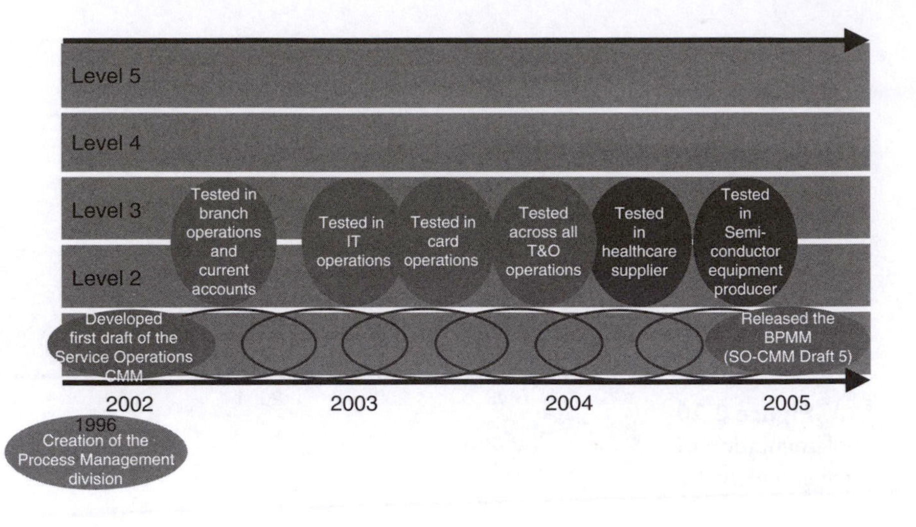

Figure 2.21 Formulation of the BPMM.

With no model to buy, the only option remaining was to build one. The organization took the bold step of contacting certain of the original CMM authors and contracting with them to write a CMM for the service operations areas – the areas of the organization essentially not covered by the CMMI, which dealt specifically with the world of projects and programmes.

The model was started in 2002 and the first version was developed and tested later that year. Because it was needed to address the challenges of all the different operational areas it was built as a generic model for use in any area of service operations. It was based directly on the architecture and principles of the CMM's, but was designed only in the staged form and not in the continuous form. It was initially referred to as the Service Operations CMM (SO-CMM), but after the owners of the CMMI declined involvement, the name was changed to the BPMM.

The model went through a number of versions with extensive testing in all the different parts of the Technology and Operations organization (Figure 2.21). The testing took the form of different types of assessments or appraisals whereby additional good practice was incorporated back into the model. The external authors tested it with other clients of their own, and it was broadened to products and services, and made generic for use in any industry or domain. It was released in 2005 as the BPMM, and in 2006 it was presented for scrutiny to the Object Management Group and has since been adopted as a standard by that body.

Lessons learned from the CMMI journey

The CMM journey had not been a smooth one. It had yielded massive results, but had also experienced huge resistance from different groups within and from outside the organization.

In the Software Development area the road to Level 4 had required much focus and effort, but given that it was a relatively small area, it had not experienced much resistance from within; those that did not want to be part of the journey had moved on elsewhere. Where resistance and discomfort had been experienced was in the areas closest to them who now had to conform to the standards set by the strict adherence to process – they had

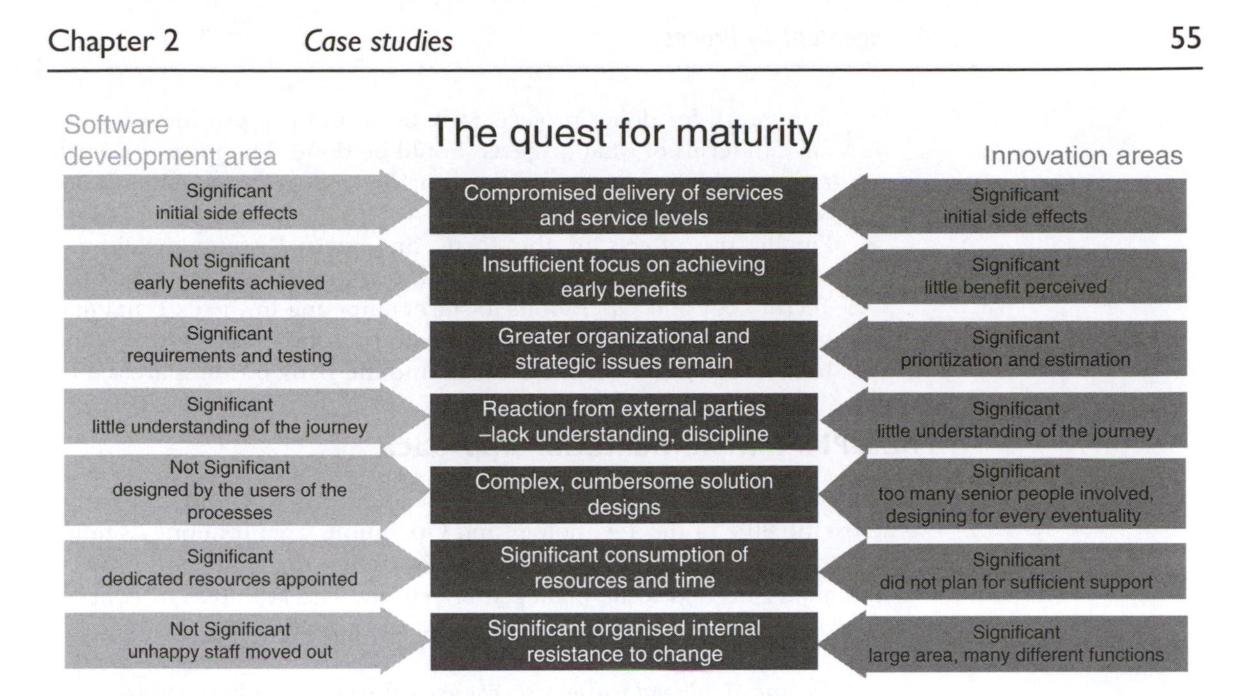

Figure 2.22
Lessons learned from the CMM journey.

neither the awareness, training, nor the discipline and thus were not properly equipped to provide their inputs to the overall process.

When these same groups became part of the CMMI journey – covering all the technology innovation processes – the training, understanding and discipline was passed on, and the problems at the interfaces between the groups began to be addressed. However, a whole new set of problems arose (Figure 2.22).

Internally:

- The size of the Innovation area and all the different functions involved meant that organized internal resistance was greater and less easily addressed – not just a matter of people moving on elsewhere.
- It also resulted in large multi-disciplinary groups being responsible for the design of the processes (a new form of PEG) and this resulted in complex cumbersome redesigned end-to-end processes – since all parties were determined to provide their full input (have their say).

Externally:

- The business side of the organization did not understand what the Innovation functions were up to and similarly felt the effects of a lack of training and discipline on their side.
- The journey also required huge effort and internal focus with the consequent reduction in the time and focus that could be, and was spent on the external relationships.
- Furthermore as good as the model was at highlighting the internal process issues for resolution, it did not look at the more strategic organizational processes. In other words the CMMI provided the

framework for doing projects well, but it did not provide any guidance in terms of what projects should be done. The project prioritization and estimation processes remained a huge point of frustration with the business areas.

- Finally, the effects of the focus on external commercialization were felt by the business. As much as the process journey allowed Technology and Operations to start competing in the external market, the massive investment in time and resources meant that somebody would have to feel the pinch and the Bank business areas did.

The BPMM implementation approach

So what could be done differently to make the journey in operations less arduous across the 80% of the Technology and Operations organization? As much as the model is a change management tool in itself – the change of introducing the model needed to be managed effectively. The key areas of concern identified from the CMMI journey were addressed as follows:

- *Organized internal resistance to change* – through familiarization.

A lot of time was spent familiarizing all the different operational functions, particularly the management level, with the journey, what it was about, what involvement it would require, how we would achieve the targets that had been set. This was done through:

- Introductory workshops with managers
- Follow-up solution workshops with managers
- The Orchestra event.

The Orchestra event was a particularly powerful introduction to the concept of process maturity. At one of the Top 1000 events, a full symphony orchestra was used to illustrate the five levels of maturity. At Level 1, the musicians were scattered individually throughout the audience each trying to play the same piece of music, but with almost no co-ordination or harmony. At Level 2, the musicians were clustered into the main instrument groupings, but the groups were still scattered throughout the audience. Each grouping played with co-ordination and some harmony, but it did little for the music as a whole. At Level 3, the musicians were gathered on the stage in their instrument groupings, the overall co-ordination and harmony was restored, but there was room for improvement. At Level 4, the musicians made use of measurement and the combination of the parts to further improve the overall performance. At Level 5, the conductor, having set the parameters for his interpretation of the piece, allowed the individual musicians to perform to their own individual brilliance. The parallel was remarkable, the result was unforgettable.

- *Consumption of resources and time* – through involvement.

It was ensured that there was sufficient representation across the operations in all the solution development activities (process design), of dedicated staff, at the right level of experience (enough to contribute, but not be distracted).

- *Complex, cumbersome solution designs* – through concentrating on simple, flexible solution designs.

A continuous focus was maintained on developing simple, standard, re-usable solutions.

- *Reaction from external parties* – through support.

Direct support was used to assist with the efforts to educate customers on the requirements and dependencies of the operation, to improve understanding and discipline.

- *Greater organizational and strategic issues remain* – through awareness.

A strong emphasis was maintained on the fact that the Maturity Model is not a roadmap for strategy development, rather a means for carrying out strategy well.

- *Insufficient focus on early benefits* – through focusing on early benefits and simple successes.

The results of the appraisals were used to identify areas where simple solutions could yield the most and immediate benefit, the greatest points of pain that would require the least amount of fixing, for example, capacity planning.

- *Compromised delivery of services and service levels* – through direct support.

Direct support was provided through

- In-house SO-CMM co-ordinators (ex-solution development team) and
- a central team of SO-CMM specialists.

The appraisals conducted across the different parts of the organization were used both to test the model and identify good practice for inclusion, as well as to identify the gaps or improvement opportunities in the different operational areas (Figure 2.23). Model specialists and selected representatives from the line areas then jointly developed standard process solutions – these were further defined for their own particular business line areas, and implemented and supported by both groups. By using a partnership of business line representation (SO-CMM co-ordinators) and process specialist, ownership of the solution was maintained in the business line, but sufficient support was made available for effective implementation.

The top issues identified and addressed

Work unit management

Numerous problems were identified in managing work units and they differed from area to area. Some of the most serious involved the imbalances between workload and the available staff. The causes of imbalance included:

- commitments being passed down to work units that they did not have the resources to meet;

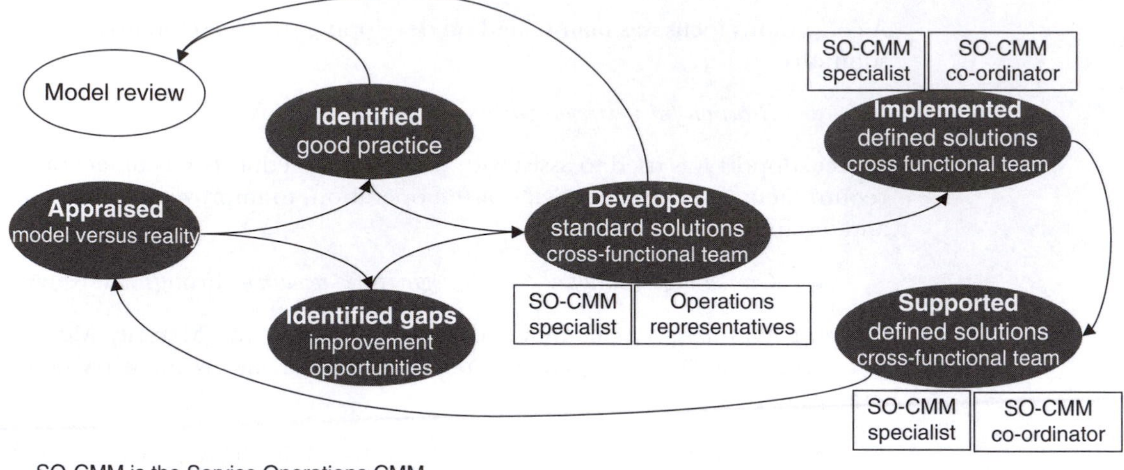

Figure 2.23
Development of standard process solutions.

- work requirements that were not sufficiently understood for estimating the resource needs;
- changes to work requirements that were not reconciled with existing work commitments;
- commitments that were not reviewed with the product or service customers;
- Service Level Agreements that were outdated or did not contain sufficient detail for estimating work effort;
- inadequate agreements for co-ordinating work among the different work groups that shared dependencies in meeting commitments;
- managers who lacked skill in estimating, planning, negotiating and monitoring workloads;
- inaccurate planning parameters or models for estimating workload;
- a lack of accurate historical data from which expected performance levels could be analysed.

The problems were addressed by developing and implementing processes for the definition and acceptance of work unit requirements, the negotiation and acceptance of work unit commitments, providing the tools and training for planning (demand, capacity, resource, skills, risk), the establishment of meaningful and measurable operating and service level agreements and effective status reporting.

Work measurement

Numerous problems were found in using measures for managing work. Again specific measurement issues differed between areas, but several problems emerged that were common across all areas. Although many measures were collected and dashboards were used for tracking progress, the measures were not being effectively used for estimating and analysing work. Some of the problems included:

- measures in capacity models were based on industry standards for transactions that did not take into account rework or work that must be performed separate from a transaction;
- measures were not analysed to understand and calibrate key factors such as rework;
- measures were not always tied to defined processes;
- measures were not fully integrated into work unit management activities.

The problems were addressed by introducing what was referred to as a 'measurement specification' for each work unit, where all measurements were defined as part of an overall measurement construct which defined the base and derived measures, the data and the reports. By understanding the relationships between the different measures, the relevance and usefulness of the different measures could be reviewed. Use was also made of the Goal/Question/Indicator/Measurement technique to ascertain the appropriateness of measures.

Process compliance

Numerous problems were found in complying with the processes defined in the process models. Process fidelity (that is, processes are performed as defined) provides the foundation for the benefits of a process culture, but the existence of process models is not sufficient to ensure process fidelity. Process fidelity represents the extent to which processes are being performed in compliance with their description. Most organizations require one, and in some cases two years to achieve process fidelity, and even CMM Level 3 organizations can experience problems in achieving process fidelity. Some of the problems affecting process fidelity that were reported include:

- process definitions that did not represent the way certified professionals found most effective for performing their work;
- inconsistency among different representations of the process;
- difficulty in understanding the case tool models;
- the size and complexity of some models;
- lack of assistance in learning how to comply with defined processes;
- process models not synchronized with related materials causing confusion;
- the frequency with which changes are released.

The problems were addressed by setting up PPQA/PPA (Process and Product (Quality) Assurance) teams to audit process conformance, to support the process users, and to identify risks and improvement opportunities – in effect to assure process and product quality.

Although many other issues were identified including end-to-end service co-ordination, configuration management, business governance and others – the above represented some of the key Level 2 issues that were identified and addressed. The response was remarkable as managers were now being helped to find solutions to their day-to-day management problems. As much as smart new technology assisted in reducing work and getting the work done more efficiently, the management problems seldom diminished and the expectations of more to be done with less continued to increase. Managers were seldom in a position to establish exactly what they should be able to do with what resources and relied mostly on experience and feel for the job. Now they could push back to the senior levels with proof of what demand could be met with what capacity, and what levels quality and risk could be anticipated – the right decisions could be made at the right level.

The journey derailed

During 2003, the organization experienced some major difficulties, not as a result of the process journey, but rather as the cumulative result of a number of negative issues which converged at a crucial time (macroeconomic, acquisition and strategic). This had a substantial financial impact upon the bank which resulted in a re-focus and a complete change in the top leadership structure, with the replacement of many of the executive, the restructure of business and operations from the horizontal integration of the old Technology and Operations division to vertically integrated product monolines and a stand-alone technology division. The process management function was chopped into pieces to support the different parts of the bank. The Bank moved into recovery mode with a stringent focus on cost control and on beating the recovery targets that had been promised to the market. The plan was three phased: fix the business, consolidate and grow. With the disaggregation of the process function, the focus on cost and a strong reaction to the immediate past, the sponsorship of the process maturity journey (CMMI/BPMM) was stopped. Process management continued in name, but not in reality as all the business line areas focused on containing their costs. Much of the process skill base was eroded as the process engineers sought new opportunities now that the priorities had changed.

Post recovery

Three years later the bank has recovered – it has met all of its recovery targets, moved from fix and consolidate back into a growth phase. The turn around has been remarkable and the organization is poised for significant further growth. The structure has remained vertically integrated and the process management function remains broken in disparate pieces.

What effect has that had on the organization? The need for business analysis and business requirements development skills never diminished. What appeared was a void which process management used to fill, which was consumed by external consultants with the consequent mish-mash of quality, no re-usability and no retention of the intellectual property. The process skills where they have remained have been stretched to the limit, the need has not diminished. Where the skills have remained, all the original intellectual property has been resurrected and re-used to great effect in terms of regaining market share. Even though there is no longer an organizationally espoused process journey – the memory of the organization has endured and the need for business process management is recognized. Where the skills in process maturity have remained, the focus has moved strongly to address the Level 3 goals, through the definition of business services and the development of the service management function, as well as the integrated process architecture and library. The PPQA/PPA process assurance function has endured and is starting to grow and support the drive to a complete process repository. In many of the operational areas the work unit management solutions have endured only because of the implicit benefit they have provided to the managers concerned, with no support of the management levels above.

The question still remains however, on how long that memory will endure and what the effect of a distributed existence will have on the process management discipline in the longer term. The Bank has embarked on a sponsored Service Oriented Architecture initiative and the lure of the new and improving BPM technologies should both focus the development of the process discipline – but what of the goal of a process culture?

There is no doubt that the journey to a process culture requires strong, central, executive sponsorship and leadership. There is also no doubt of the organizational benefit that can be attained through that journey. At present the organization is striving towards achieving a common culture, it is also investing millions in developing management skills, and it is starting to develop a strong focus on services, albeit in the technology area. All these are addressed directly by the BPMM and the CMMI, but the memory of the past is harsh and it will be interesting to see how the latest journey unfolds.

Conclusion

The benefits of business process management with or without the application of technology are well proven and have been successfully harvested within the Bank for the last decade.

To deliver the benefits of new technologies the organization required a common and collective understanding of what happens in the different parts of the business (an end-to-end view); it needed documented business processes; with the rules, activities, dependencies, inputs and outputs of all the multiple back-office functions specified.

Similarly to improve performance in the operational areas – as a result of the increased focus on improved customer service, lower costs, improved turn-around times, better quality and reduced risk – standard and aligned processes were required.

However, co-ordinating, maintaining and managing such a journey is a complex challenge and that is where the value of the maturity models, the CMMI and BPMM, became paramount. They provided the roadmap for the process improvement journey, they provided the balance between the very necessary focus on individual business re-engineering efforts, the continuous improvement perspective and the management processes which had for so long only been a secondary focus. Local improvement can be achieved with relatively little effort, but how does an organization create an enduring organization-wide improvement effort. That requires a common culture, and the maturity models provide the reference framework and the means to attaining that common process culture.

Many software development organizations around the world continue to use the CMMI to reach levels of performance not yet considered by other service organizations. The BPMM was developed to apply those concepts and that approach to the greater organization and has been used to great effect in the Bank, notwithstanding the short period and the turbulence the organization has experienced.

Business process management views processes as capabilities that must be managed as enterprise assets – that means planning investment in them, developing and improving them and governing their performance. Business processes are no longer handled in isolation from each other – they are now looked at as a set of enterprise capabilities, and as such the set must be dealt with as a whole. They must be governed; remain true to statutory and regulatory requirements, and to organizational policy; be synchronized to the strategic intent; use resources wisely; ensure accountability; maintain alignment and monitor and control initiatives.

To do all this requires an integrating framework and a governance mechanism based on sound, yet fundamentally different management principles.

Case study – Aveant Home Care organization

This case study is about an organization called Aveant which provides Home Care services in The Netherlands.

Background

In The Netherlands the health care systems is undergoing fundamental change, especially in the home care sections. Main developments include:

- changing legislative regulations
- an increasing open market
- more competition on price and quality
- consolidation of the various health care and home care players
- increasing demand for health and home care services (with an ageing population)
- difficulties in recruiting staff.

Aveant is an organization that is the recent amalgamation of two organizations – Cascade and 'Thuiszorg Stad Utrecht'. It has about 3,000 employees with a turnover in 2005 of €85 million.

Business challenge

Due to the increasing demand and scarce resources the organization was facing problems in the areas of:

- Increasing processing of client requests – some requests took up to seven weeks for completion and the waiting lists were increasing.

- Service quality came under increasing pressure – there was much duplication of information, including re-typing of timesheets, leading to high error rates.
- Cost of service provision and support started to increase – the government wanted to force a more market-oriented approach and they were forcing this to occur by reducing subsidies – so there was an increasing need to improve productivity.
- The organization was struggling with insufficient agility to enable it to deal with the changes in the marketplace.
- Management of the business processes became more and more difficult – for example, business planning was completed on a large whiteboard with post-its notes and pins. Needless to say, this simple system was labor intensive and prone to errors.
- While the general demand for home care services was increasing, there was also a growing expansion in the number of services required by individual clients.

The organization needed to ensure that its scarce resources were being focused on doing the 'right things right'.

Review project

Approach

To overcome these marketplace and organizational challenges the CEO initiated a project and took personal charge.

Business alignment

The CEO and his direct reports formulated a five year plan with the following key elements:

- To work closely with partners to provide an integrated home care service – home care, well being and housing.
- To recognize the differing client needs and to provide adequate and appropriate services.
- To have reliable, automated and integrated planning and logistics.
- To improve communication with the employees and partners.
- To record and manage knowledge, expertise and embed it in the organization.

Improve process thinking

The employees and partners were educated in the power of business processes and were encouraged to think in these terms, rather than the more traditional functional perspective. They were required to think in terms of:

- Better customer services
- Focus is on the performance of the end-to-end business processes and not on individual tasks or activities

- Market-orientation should replace production-orientation
- Management supports and facilities instead of controlling and enforcing
- Encouraging employee involvement in the end-to-end business process, rather than them just being restricted to their own activities
- Automation handled the standard mundane transactions and exceptions were passed to specialists to handle
- Improved employee satisfaction
- Reduction in the amount of paper produced
- Reduction and controlling the throughput time
- Improving the insight and manageability of the business processes
- Ability to obtain and provide real-time status information of processes, clients and patients
- Realizing that much of the current work was either corrections of previous work incorrectly completed or work that was non-value add.

The business processes were positioned at the centre of all project developments (Figure 2.24). This provided management, the project team and staff with a common foundation and understanding.

Project findings

During the initial phase of the project various business processes were reviewed and assessed on their potential for automation and improvement, including obtaining process metrics. The idea was to identify low-hanging fruit to proof the benefits of the process improvement and automation. Key indicators that business processes may be a candidate for automation and improvement included:

- The degree of standardization of the process – measured by the number of exceptions
- The length of the process – measured by the number of steps and the complexity of the process
- The volume of the work – measured by the number of instances
- Work assignments – measured by the number of roles and people involved.

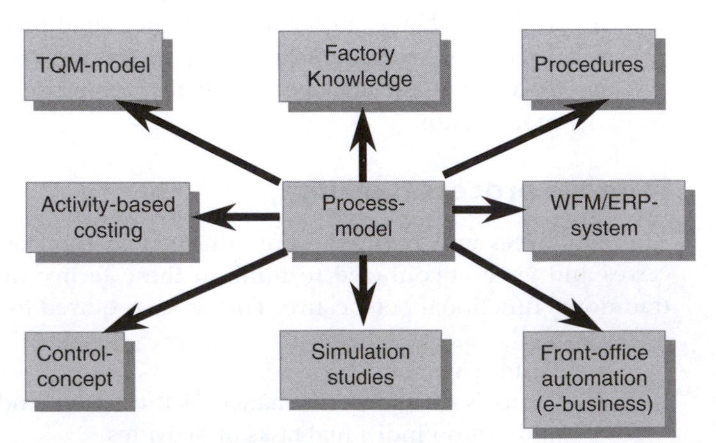

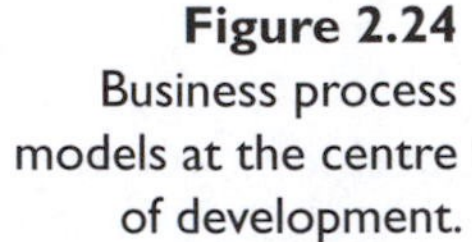

Figure 2.24 Business process models at the centre of development.

	Standardization	Length and number of steps	Number of cases	Work assignment	Amount of information
Customer Service (indications)	+	0	+	0	0
New clients	+	+	+	+	+
Execution of care	–	–	+	+	0
Planning	+	–	+	–	0
Administration	+	–	+	–	–
Course administration	+	0	(+)	–	0
Customer services: changes	+	–	+	+	–
Time registration	+	–	+	+	0
Salary-administration	+	–	+	–	+
Personal Administration	+	+	+	+	+
Sick leave administration	+	+	+	+	0
Purchase: placing orders	+	0	+	?	+
Billing	+	–	+	0	+
Purchase: financial admin	+	+	+	+	+

Suitable for workflow

Potentially suitable for workflow

Figure 2.25
Process analysis.

The analysis shown in Figure 2.25 was completed:

Business pilot: Starting small

The first project that was undertaken was small scale and related to the Human Resources (HR) processes, such as salary, sick leave administration, recruitment and releasing of staff. These processes were labor intense and had a high error rate. They were relatively small processes with a limited number of people involved, however, they impacted nearly everyone in the organization.

Approach

A multi-disciplinary project team was established to ensure that the processes were reviewed from all the relevant angles and disciplines. Employees were encouraged to contribute to the discussion and provide suggestions.

The bottlenecks in the processes were identified and the business rules for the processes were critically reviewed. For example, previously every manager could determine on which day a new employee would start. This was changed to have a common start day of Monday. This simplified the process and allowed for substantial streamlining without much disruption or discomfort to the organization.

The current processes were modeled using a whiteboard with magnets for process symbols. This allowed flexibility and the ability for everyone to participate and make modifications themselves.

Then the processes were critically reviewed. What do they want to achieve? Does the organization need to do these process steps? Is the process being executed smartly? The streamlined processes were then entered in the computer.

Automation played a key role in the streamlining of the processes. The introduction of workflow provided better oversight and flexibility for management and reduced errors and throughput time. All the work was assigned via email. Follow-up messages were send in cases where the work was not progressing sufficiently. It also allowed the organization to gain insight in its backlog and the outstanding workload of staff that called in sick, and enabled management could take appropriate action.

In the development of the workflow systems, the organization took a very pragmatic approach towards processing exceptions. Rather than spend 80% of the development effort on less than 20% of the cases, they implemented a basic workflow with an 'exception button' on every screen. This button would override the normal business processing rules.

Staff in the quality assurance team assessed each exception. Firstly, they determine if it was really an exception. If not, the case would go back into the workflow and the relevant employee would be informed accordingly. If it was an exception the quality assurance staff would process the case.

The benefits of this approach were

- Reduced programming and testing required to deal with all current and future exceptions
- Workflow was simple and easy for people to be trained
- Experienced staff dealt with the exceptional cases, which are more complex and time-consuming to process.

Results

The results were very encouraging. Due to the streamlining and improvement of the business processes:

- the quality of the work improved
- the number of errors reduced significantly
- administrative staff could be reduced with 50% in number – in a marketplace where it is difficult to obtain qualified and motivated resources, this reduction had a significant impact upon the organization.

As employees witnessed the improvements this approach was making to the organization they started to look for more opportunities.

Continuation

The success of the initial project was used to initiate a larger programme where other areas were also assessed for opportunities for improvement.

Approach

The further continuation of the programme is actually seen as a process itself. The earlier success gave the employees more confidence to start tackling more complex and larger processes. A noticeable shift was made from just supporting processes to processes being the primary focus.

The organization realized that by initially selecting the 'low-hanging fruit' they were able to provide staff with confidence and competency in the approach and the commitment and perseverance to tackle the challenges ahead.

The project continued to collect metrics on the business processes identified for improvement. This provided a proper baseline for future comparative purposes and ensured that decisions on which processes should be improved were based on hard facts and not just guesswork or perceptions.

Many of the business processes were facing problems, as much of the required data was missing or the logistics were not aligned. All improved processes were clearly defined and all 'noise' and non-value add was removed, so that there was no ambiguity left in them. Efforts were made to train the staff to ensure that the right questions were asked of customers and that the optimum solution was selected.

The EFQM model was used for the modeling of the business processes. All process models were directly entered in a quality system to ensure that they were completely aligned.

Effort was made to empower the employees so they had more influence over the process and that they could make more decisions themselves – in line with a clearly defined decision-making framework. The employees also obtained a better insight into the business processes and key bottlenecks. Employees were encouraged to provide suggestions, recommendations and to identify opportunities for improvement.

Effort was made to ensure that every business application system was properly developed and aligned with the business processes and that it was reliable and robust enough to support the staff in their work.

During the course of the programme more and more business application systems were included in the scope of work, such as the financial administration. PDA's were introduced to ensure that everyone had access to the most up-to-date information. The PDA's were positioned as a way to empower the employees rather than a piece of technology that is nice to have.

The CEO, Evert Mulder, played a prominent role in the project: supporting the scoping of the project, the innovation of the business processes and the implementation. Much time and effort was initially spent on people change management: explaining the purpose of the programme over and over again; relaying this message in different ways and on different occasions.

The people change management aspects did not only relate to the processing staff but also to the middle managers, especially as the managers number of direct reports diminished.

The CEO was involved in most of the communication, especially as at the beginning of the programme as many different forces impacted the programme. A consistent message from the leadership proved to be critical to success. Once several projects were implemented successfully and people were involved, committed and capable, there was less need for the CEO to be as closely involved as he was at the initial stages. In fact, an informal group of enthusiastic employees was formed who were keen to improve the processes; raise organization-wide process awareness and commitment; and ensure the leverage of the increasing momentum. This group was instrumental in getting the projects launched and completed successfully. They created a 'can-do' attitude that gave confidence and energy to other employees.

The organizations experience has been that subsequent implementations of business process improvements were progressively easier, especially as employees were informed and committed to these projects. Later projects did not need to be sold to employees as to their purpose and need. Staff was continually kept informed about the portfolio of projects and could do online courses to be trained in the new way of working.

Parallel to the process improvement projects was the creation and assignment of process owners and a process governance structure. This ensured that the right people were involved in process improvement and that there were mechanisms to monitor progress, impact and the feasibility of these improvements.

Results

To date five projects have been completed and each of them have similar results:

1. significantly streamlined business processes
2. reduced error rates
3. 50% reduction of FTE staff involved in the process
4. a reduction in the dependency of paper
5. throughput time has been reduced up to 80% – in the most extreme case it was from seven weeks to four days.

The employees were extremely pleased with the results as it eliminated a lot of unnecessary 'noise' in the processes and they could concentrate on their actual job, rather than correcting errors or chasing other people to complete their work correctly.

The organization recently faced a big challenge with a merger with another organization. This is typically a situation where much of the progress achieved in one organization can be difficult to continue to progress and translate to the other.

However, in this situation the business process management approach and methods were accepted by the other organization. Both organizations found that the process management approach was an excellent way to get the best out of each others business processes.

Within three months the merged organization had implemented the process management approach throughout its organization.

During our interview with the CEO, Evert Mulder, he stated:

> It is amazing to see that although business process management can provide such successful results that there are so few companies and leaders willing (or daring) to truly support and implement business process improvement initiatives

Authors comments

Leadership

This case study confirms, yet again, that organizations that have successfully implemented consecutive business process improvement initiatives have a leader that has the courage and dedication to initiate and support a programme. They have a clear vision of what they wish to achieve and understand that this is the best way to achieve it.

Leadership is not just required at the initiation of a project, but also during implementation. In this organization the CEO ensured that the project was focusing its attention on a pragmatic solution starting with the low-hanging fruit. In addition, the leader was able to inspire a group of people to actively participate, create a '*can-do*' attitude and to involve their colleagues.

Process governance

Process governance was ensured through the appointment of process owners. These process owners were involved in the decision-making about improvements and were responsible that the processes were compliant with legal requirements. The CEO now has a next level of management to whom he can delegate the accountability for the business processes.

The quality assurance for the business processes is becoming embedded in the process itself, especially through the control over the exceptions.

Process execution

The workflow application provided information to the process owners about the progress and performance of the processes. Bottlenecks were easily highlighted and any backlog of transactions can be reassigned.

The process execution improved significantly as the employees became more empowered. This has resulted in an increase in employee satisfaction.

Strategic alignment

The link to the strategy was a critical success factor as it provided the employees with the burning platform on why the business processes must be improved. The organization was looking at a paradigm shift rather than just a gradual change, or even worse a change dictated by external market and regulatory forces. The strategy also provided the necessary inspiration to think outside the box and look for significant and sustainable solutions.

The choice of a model to link business processes with strategy is not as important as to making sure that the model is applied consistently.

Project execution

Projects were well-planned so that the organization moved gradually from 'pilot' projects to 'in the driver's seat' projects and have finally reached a

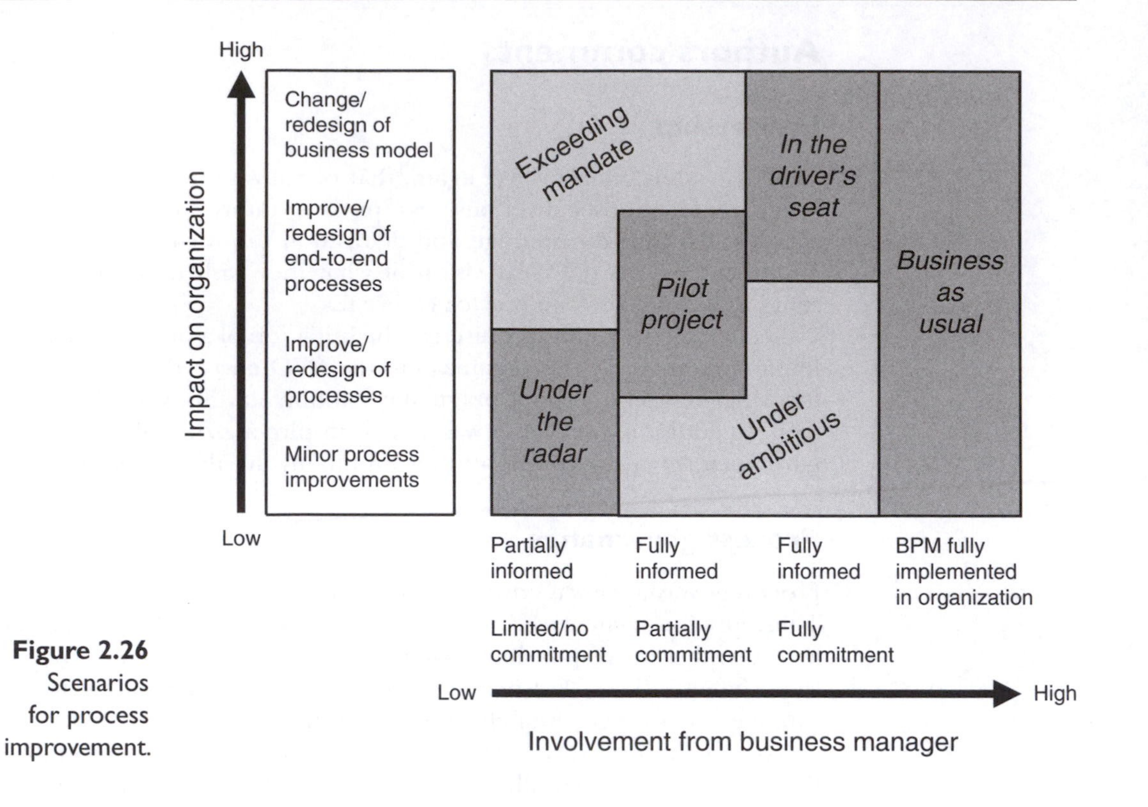

Figure 2.26 Scenarios for process improvement.

state of 'business as usual' (in line with our growth model (Jeston and Nelis, 2008) shown in Figure 2.26).

This approach ensured that the organization was ready to deal with business process changes that had a large impact on the organization, its customers and its partners.

People change management

Communication has been a critical success factor. Communication was not just limited to a newsletter or slogans on a poster. Key was that the CEO himself publicly supported the project.

Sufficient attention was given to the impact of the changes on the role and position of the people impacted.

We share the amazement of the CEO that given the success stories on having a business process-focus can bring to an organization, that there are still so many organizations who either 'do not get it' or do not have the courage to take action.

Case study

This case study is about an international bank which ranks among the largest in Europe.

Business challenge

The organization initiated a global programme four years ago to improve the services it provided to customers and to streamline the back-end banking operations. The program was aimed at consolidate the branch offices and centralize back-office operations. In addition, for specific areas such as lending, specialist branch offices were created. The ultimate objective was to lower overhead costs and be more competitive in the marketplace.

The European Service Organization faced the following key challenge:

- The current processes were highly manual based
- There were a high number of processes and process variances within each, which made it difficult to control and manage them
- Operating costs were increasing which was creating competitive pressure.

The business process ownership was transferred to the business line managers to ensure that business process improvement and management were more closely aligned with the actual execution of the process. Once the business was responsible for their processes they were keen to improve move to improving and managing them better.

A specific project was initiated with the aim of streamlining business processes and reducing costs. The key drivers were

- Rationalizing the individual processes by reducing the steps.
- Digitalization of the information which enabled an increase in throughput time and the ability to distribute work better.
- Off-shoring of simple activities, leveraging the workflow and imaging activities.
- Ability to track and trace the information flow for both management and clients.

Review project

Discovery

The executive manager took a leading role in this project and, together with his management team, identified about 20 processes that would be reviewed as potential areas for improvement.

The project reviewed all these processes and assessed whether there was scope for streamlining. Six Sigma methodologies were used to obtain key information about the performance of the processes and the scope for improvement.

The project found that for reviewing large end-to-end processes, the Six Sigma methodologies were quite cumbersome and time-consuming. A pre-assessment had to be made as to whether or not an improvement opportunity warranted the level of energy and resources required to perform the Six Sigma detailed examination.

Each process was assessed on the basis of its potential for:

- streamlining
- automation and digitalization (including improved tracking and tracing)
- off-shoring.

The executive manager and his team ensured that this was in alignment with the corporate and business unit strategy.

For each process the following questions were asked:

- Does the process need to be performed in this way, or are there better ways of streamlining it?
- What is the preferred way of interaction with the customers?
- How can we make it more transparent?
- How can we improve tracking and tracing?
- Can we digitalize it?
- Can we offshore it?
- Can we segment the process by customer to improve customer servicing?

Off-shoring

The management decided that the simple straight through processes would be outsourced and processed in India. All work that needed to be performed in India had to be entered into the imaging system and the workflow application.

It was interesting to note that initially, off-shoring was not included in the scope of the project. However, once the opportunity arose to offshore certain activities, it was included in the scope of the project and quite seamlessly added to the process review framework. The organization stated that:

> off-shoring required us to create a few additional trays in the workflow management system, but otherwise off-shoring merged quite well with the automation.

Metrics analysis

The aim was to offshore about 70% of the work to India. The main criteria was that the customer should not notice a difference in the level or type of service provided. It was envisaged to keep about 30% of the work in Europe. The key reasons for this were:

- Most communication with the customer was in the native language, hence the need to keep on providing local support for customer interactions.
- Contingency – the organization wanted to have a fall-back scenario in case of any problems with the Indian processing centre (including the communications link of data going to that centre).
- Escalated issues or complex processes were still completed from The Netherlands, leveraging the years of experience and expertise of the European team.

- Close contact to the business process – so that the organization can provide direct feedback for product development. It was considered important to keep the feedback loops open as they provide valuable information on the way to implement new projects and/or enhancements.

Some of the business processes had been in place for 30 years and were being transformed within a matter of weeks. A conscious decision was made to provide European employees with an opportunity to train their Indian counterparts. Although many of them were not fully qualified to provide training, they were able to provide a wealth of knowledge, skills, expertise and experience to the Indian team.

Management was impressed with the high level of education in India. In Europe the majority of staff had basic schooling at high school level plus a few years of courses. In India, the majority of the people had a university degree.

Process awareness

The executive management wanted to change the culture of people and management. The managers were sent for Lean Six Sigma training so that they would understand the importance of business processes and the need to manage and improve them with the knowledge of metrics. This training will be extended to subject matter expertise in the future. This ensured that everyone had a consistent approach to process analysis and improvement.

Management information

To achieve continuous improvement the management wanted to have consistent, reliable and timely management information. Up to this point, each department had their own management information system, with varying definitions, timeliness and quality. It was envisaged that each department would link to a centrally provided solution.

Management information was developed at a more granular level than previously. Key management information was projected on a billboard on the work floor. People were also able to see at the coffee machine how they were performing.

Initially there was resistance to the openness of the management information. Now everyone is asking questions such as:

- Why is this month's performance lower than last month?
- Why are we doing the process like this?
- How can we improve things?

Leadership

During the transformation the executive manager spend 60% of his time on people change management, especially with a great effort in communication and ensuring that all initiatives were still aligned with each other, with the overall strategy and the business case. The remainder of his time was spent on business as usual activities.

A critical success factor was the leadership of the executive manager and his management team. They initiated change, demanded change, encouraged change and they 'walked the talk'. They became more pro-active and no longer accepted certain process shortcomings and continuously looked for better solutions that were more sustainable.

These types of projects take a significant period of time to complete and realize business benefits. The programme of work has been going for the last 18 months. It is considered critical that management keep focused on the required outcomes and not be distracted with other, less important, issues. The vision was formulated at the beginning of the project and it was considered essential that it was executed well with no distractions.

People change management

Communications played a crucial role in the project, especially as several hundred employees faced redundancy. At these moments leaders have basically two choices: ignore the eventual redundancies or be very open about it. They chose the latter as they wanted to be honest and open with employees, as they have a right to know what is happening to their jobs.

The purpose, business needs and approach for the project and its outcomes were repeatedly communicated, in a planned and consistent way so that there was no ambiguity. This also provided employee confidence in the outcome of this project.

People were openly and honestly informed about the plans. This allowed management to outline their plans at an early stage and provide coaching to employees of the options available moving forward. The selection of the people to stay was based on their performance as well as their ability to change to the new way of working – flexibility became a key characteristic.

Interestingly enough, several of the employees that were made redundant were actually keen to provide training sessions to the people who were taking over their role.

Results

- Processes have a significantly faster throughput time (for some processes the throughput time has been reduced by more than 50%).
- The organization is now in the process of phasing out 300 employees and have recruited about 100–110 people in India to do their work.
- The organization has become more agile as more employees question the current practices and provide suggestions and ideas on how to improve them.
- The organization is on track to have 80% of their work digitalized, providing them the planned benefits of remote work as well as faster throughput times.

The authors analysis

Leadership

This case again highlights the importance of management buy-in and actually driving the initiatives. It is also critical to ensure consistency throughout the management team. Having just one person driving the initiatives is not sufficient to deliver the required change.

People change management

The organization, just like any other for-profit organization, is facing a challenging increasingly competitive environment and step-changes followed with continuous improvement are required to deal with them. This project, although painful for the people made redundant, provided more certainty for the people remaining, as the organization as a whole will be more competitive. This case also confirms the importance of open and frank communication to the employees.

Process governance

The process governance and ownership assisted with a more ideal situation and drove the change. Often process governance is too distanced from the actual execution which can result in too much bureaucracy and less incentives to actually make changes.

Automation

The automation provided the potential to streamline processes and to perform the off-shoring. However, the biggest challenge was more on the process and people-side than system issues. This is certainly consistent with our experience.

Strategy

The formulation of a clear vision, mission and strategy provided this project with the much needed guidance to perform the project. The consistency of the strategy allowed the management to remain focused.

Manage the business case

The introduction of the off-shore option was not included in the original scope. However, it was found that it could contribute significantly to the overall project objectives. Rather than lock the scope of the project and ignore the option, the project team redefined the scope and included it in the project through proper project change control. The off-shoring significantly contributed towards the success of the project.

Case study: Air Products and Chemicals, Inc.

The following case study has been researched and written as a result of an interview with George Diehl and reference to the American Productivity and Quality Councils APCI case study (2005). George Diehl was the global director of the Process Management Centre of Excellence and supply chain and process management education lead for the six years between 2000 and 2007. He has since retired and now is a Business Fellow at Villanova University, Villanova Pa. George was gracious enough to be interviewed and then review this case study to ensure its accuracy and completeness. The information herein largely applies to activities witin the timeframe of George's process management tenure. We would also like to thank the American Productivity and Quality Council (APQC) for its permission to use diagrams, information and text from its 2005 case study article.

Background

This is the story of a profitable multi-billion dollar geographically diverse organization with 22,000 employees worldwide. Air Products operates in more than 40 countries around the world, with half its customers outside the USA. Air Products sells gases, chemicals, equipment and serves customers in technology, energy, health care and industrial markets.

It is the story about how one man's vision for an organization can substantially change it and position it for greater growth and profitability. One person alone, however, is obviously not enough. In the end, it is the team approach and team actions that achieve the outcomes. However, without the vision, drive and consistent demonstrable support of the person at the head of the organization, the outcomes are either significantly harder to achieve, or not achieved at all.

At Air Products a new Chairman/CEO was appointed in 2000. John P. Jones III had been with the organization for 28 years before his appointment as the organization leader. While the organization was doing well, Jones saw inefficiencies and the typical silo view of the world – 'my' business, 'my' function. From a process perspective, this had the predictable outcomes of unique processes for the same business activities by function and geographical region. Standardization was perceived as not being possible because 'we are different, unique'.

The following are the questions asked of George Diehl.

While the leadership of the Chairman was obvious from the background, how did Jones get the organization ready for the changes he perceived as needed, and how did he get executive buy in?

In order to start the business transformation that Jones saw as necessary, he created a unifying vision called 'Deliver the Difference'. In order to galvanize support and provide a detailed way forward for this diverse organization, he created a set of guiding principles and a view of the new working environment (this can be seen in Figure 2.27).

Figure 2.27 was purposely designed so that each of the elements would be distinct, easy to read and retain.

This not only provided the vision for the organization, but also the yardstick from which to make decisions. This document became the logo or

Deliver the Difference

Our Guiding Values

Accountability
Each one of us feeling it's up to me.

Innovation
It's cherishing new ideas and translating them into actions.

Integrity
It's behaving ethically and being true to our words.

Respect
It's teams, achieving their full potential through the contributions of each individual.

Safety, Health, and the Environment
It's responsibly caring for each other, our communities, and the global environment.

Our Working Environment

Our competitive global marketplace requires a working environment which visibly demonstrates our commitment to our people. We will foster trust through open communication and consistent actions. We will share a collective understanding of our company's success and, with a sense of urgency, combine our best efforts to achieve it. We will nurture ownership behavior through clear accountabilities, recognition, and rewards. We will promote new ideas and new ways of thinking. We will value diversity and insist on an inclusive culture. We will enable Air Products' people to contribute to their full potential.

The Air Products Difference

We will be the best company to work for, the best company to buy from, and the best company to invest in. We will do this through –

Our People
... their understanding, integrity, passion, and individuality

Our Customer Relationships
... built on customer understanding, innovation, and meeting commitments

Our Shareholder Support
... becoming shareholder-friendly by meeting our targets for return on capital and growth

Our One Company Focus

We will create shareholder, customer, and employee value through market, operational, and corporate leadership.

To win, we align and agree to:

Change
- Visibly value our people in a positive work environment.

Portfolio Management
- Continuously improve our return on capital and manage our portfolio.

Growth
- Create growth through innovation and the creation of superior products and services for customers.

Work Process
- Reduce our costs through work process simplification.

Our One Company Commitment

Means understanding and listening to our **customers** as One Company by:

- Taking the best of the best and bringing it to our customers faster.
- Providing value for our global businesses through one infrastructure.
- Simplifying and standardizing global work processes.
- Globally uniting by sharing our knowledge across regions, businesses, and groups.

(21550)

Figure 2.27
Deliver the Difference.

symbol for the organization intent and people knew its contents and understood their roles. It was also displayed throughout the organization, everywhere.

Once Jones created and communicated his Deliver the Difference guidelines, how did he start to execute the changes that are outlined in this visionary document?

The first thing he did was to identify four corporate-wide business transformation initiatives that were needed to create and deliver the change desired. These four were

1. *Growth* – there was a need to identify and reinvest in growth areas. For example, R&D investment funds were redirected 75% to the identified growth 'platform' areas and 25% to the 'core' areas;
2. *Portfolio management* – this was about the continuous improvement on the return on capital and achieving the double digit growth targets established. It was about truly understanding each of our businesses and categorizing them as growth; core businesses; or businesses that needed to be restructured;
3. *Business processes* – it was recognized that there were inefficiencies and ineffective business processes across the company and thus a need to reduce costs and improve customer services by simplification and standardization;
4. *Change* – this initiative addressed the behavioral changes in the working environment that were needed.

These four areas were considered a significant challenge, and opportunity, because of the organization's geographical spread and diversity of products and services.

The next thing he did, to demonstrate his commitment to achieving the desired results, was to pull out four senior executives into full-time roles to head up each of these four areas. Together they formed the Programme Management Office (PMO).

Jones also recognized an additional critical 'piece of the puzzle'. Given the existing legacy business application systems environment, it was also decided to take the huge step of introducing an enterprise resource planning system by way of a single instance implementation of SAP within the global organization. Why a single instance? In a word – globalization. Air Products was in the process of moving to a global business and functional organization and our senior management understood the ever-increasing importance of consistent documentation and visibility across geographies and global product lines, the convergence of internal work processes and the need for greater speed in deploying process improvements to all of our customers, anywhere in the world. Rather than deploy in just one region or across only some product lines, as some companies have done, we set a course to complete the implementation worldwide by the end of 2007.

To further support the process-focused approach that Air Products was establishing, we learned that Deloitte Consulting research and experience in implementing SAP applications, revealed that 'companies that succeed in SAP implementations do so because they see the project as a *process* implementation, whereas companies that fail see it as an IT tool implementation'

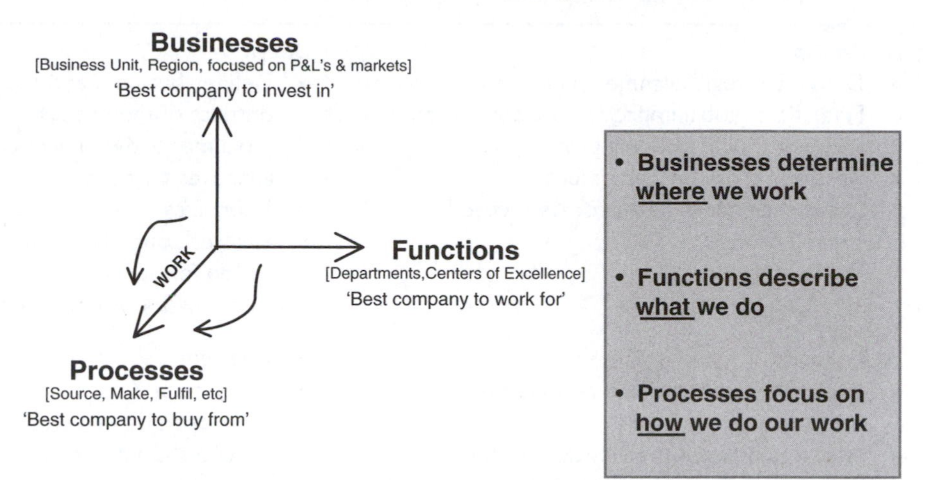

Figure 2.28 Three dimensions of management.

(http:www.deloitte.com retrieved February 2005). So the SAP implementation became a series of process enabled activities, led by senior business managers.

The company also wanted to ensure that the SAP and process work was focused on the customer and not an internally focused effort. Before proceeding with the implementation, Air Products launched a customer loyalty process, survey and a scorecard to measure customer feedback. This provided the organization with a customer baseline position and view of the organization to measure future activities against.

The organization had already established two dimensions of its management model: business units and functions. It now added a third dimension: processes. Figure 2.28 shows the three management dimensions. Executive Process Owner roles were assigned to Vice President level individuals to own the cross-functional performance of these processes across all business units.

George, the area we would particularly like to concentrate on, is the third area of Business Processes and specifically the establishment and workings of the BPM Centre of Excellence (BPM CoE). What was the primary role of the BPM CoE, how was it established and how did it fit into the business outcomes?

The Corporate Controller, who is now our Chief Financial Officer (CFO), was appointed as the full-time executive placed in charge of the business process area. The first thing he did was to establish three centres of excellence:

- Business process management
- Knowledge management
- Continuous improvement (merging Lean and Six Sigma).

I was to head up the BPM Centre of Excellence (BPM CoE). I was given the brief of not only determining how I would like to establish the BPM CoE, but of also ensuring that I and my team integrated with the SAP implementation team as well as the knowledge management and continuous

Table 2.3
Process Executive responsibilities

Leadership:	*Design:*
• Drives strategic alignment and customer focus • Prioritizes global improvement opportunities through annual planning process • Resolves cross-process issues • Leads the change to a process-focused organization	• Defines business and customer inputs and outputs of the process • Documents the process activities and approves changes • Prioritizes enterprise process IT spending • Ensures controls are in place, validated and tested for accurate financial reporting (SO_X) • Audits work practice compliance
Performance:	*Improvement:*
• Implements metrics and reports process performance • Achieves process metrics targets and goals • Prioritizes performance gaps/shares successes • Provides adequate process resources • Monitors data quality	• Analyses process performance gaps • Develops plans to close gaps • Executes Continuous Improvement projects across business units • Benchmarks and adopts Best Practices • Fosters new Continuous Improvement ideas

improvement teams. This integration also extended to always of ensuring that we also integrated with any other appropriate process activities being conducted within the organization, for example, the Sarbanes Oxley project.

The first thing we decided to do was to recommend and put forward the desired process governance structure. We prioritized seven of the company's 13 global processes based on those that SAP enabled, and requested that Jones appoint seven senior vice presidents as Process Executives who were also the functional heads of major departments. Their responsibilities are shown in Table 2.3.

In addition, the company transferred all IT spending from the business units to the Process Executives. This transfer recognized the idea that IT applications enable processes, and that these decisions are inter-related.

As these Process Executives had senior functional roles and responsibilities within the organization, it was necessary to support them with seven full-time Process Managers.

The initial structure of the BPM CoE was:

- the BPM CoE Director
- the seven process managers
- an administration person
- a couple of generalists, for example, change management.

One of the first things the BPM CoE did was to create a set of standard approaches. Things like an approach to assist the global process management teams in determining how to pick their first set of metrics and align them with the strategic objectives. These became the standard way of measuring performance after agreement with the Process Executives.

The BPM CoE, in co-operation with the SAP implementation team, also established a set of fundamental process principals, including:

- business processes will be simple, standardized, and global; enabled by a single instance of SAP and governed through a global process board.
- Customer responsiveness, operational efficiency and cycle times are all important.
- Value creation will be focused on the enterprise versus any individual business.
- Collaboration becomes the norm.

The BPM CoE also conducted many education sessions explaining what process management and process ownership meant. We also trained staff in how to establish the measures, understand the gaps between the future desired state and the current environment and the 'how' part of process improvement.

For example, we offered employees a training class called 'Introduction to Process Management.' Course facilitators began the classes by discussing the meaning of processes and the attributes of a process-focused organization. Participants rally around one simple, common definition of what a work process is: 'an organized group of related activities that work together to create value for the customer'. A process has to be organized (that is designed and then documented). It has to work with other processes, which requires cross-functional roles. And it needs to deliver something of value to the company by contributing to being the best place to work for, the best company to buy from or the best company to invest in. It is end-to-end work, not piecemeal. Facilitators make a business case for process management and explain why it is important to the company.

The course teaches that process thinking is a higher view of business. Through process thinking, employees can have a better understanding of how the entire business is operating and see the company from the eyes of the customer, not just from the organization chart.

The difficulty within most organizations is how to get functional buy-in towards becoming process-focused. How did you go about getting this buy-in?
In our organization the answer was easy and obviously the best method. The Chairman (Jones) led the charge and was fully and totally committed to the vision he established.

When he created and wrote the 'Deliver the Difference' guiding principals mentioned earlier, these were not lip service; they were lived tenets, and people needed to get on board. Where significant issues arose between senior managers or functional positions, these guiding principals were used to guide and break the deadlock.

An example of Jones' commitment to this process was demonstrated by his attendance at the monthly SAP project board meetings – he did not miss a meeting for four years – it was simply *that* important to him and this message certainly 'got out' into the organization and management.

Another example was that once a process executive gained an understanding of the benefits to be gained from a process-focused approach to their business and had 'bought in' to the process journey, we nurtured the executive's commitment by having them speak at conferences and explain, to others, the progress that Air Products and they had made with a process-focused organization. These conferences also enabled the executives to network and learn what other organizations were doing.

We also developed a corporate version of the dashboard and the Chairman looked at it regularly. He would use it to monitor the business performance and as a means of then following up with management and asking questions.

You mentioned a Global Process Board how did this work?

We had identified 13 global processes, each of which was led by a global process management team. The executive process owners from each global process management team met regularly as a global process board. This group had responsibility for documenting process design, identifying key performance indicators, training, managing best practices and resolving issues across the global process management teams, all linked to the SAP implementation. Later, this process board evolved to the Global Supply Chain Board as we recognized the need to include business representatives and a focus on cross-process issues. The 13 global processes are shown in Figure 2.29.

The organization liked the one word descriptions from the Supply Chain Operations Reference (SCOR) model. Figure 2.29 shows the seven customer-facing processes, which are the processes of the supply chain plus two additional processes. *Innovate*, the new-product development process was originally known as *create and improve offerings*, and begins with new ideas and ends with a new marketplace offering. *Sell*, was originally known as *find, win and retain customers*, starts with a lead, through proposal and contract and ends when a customer is on-stream. Once on-stream, the customer is served by the supply chain processes.

As each of these process areas was very large, we found that it was too big to be managed globally by one person, so the process was divided into three or four sub-processes owned by global process owners.

You also mentioned a corporate dashboard that reflected the performance of the processes. How did process measurement work in the organization?

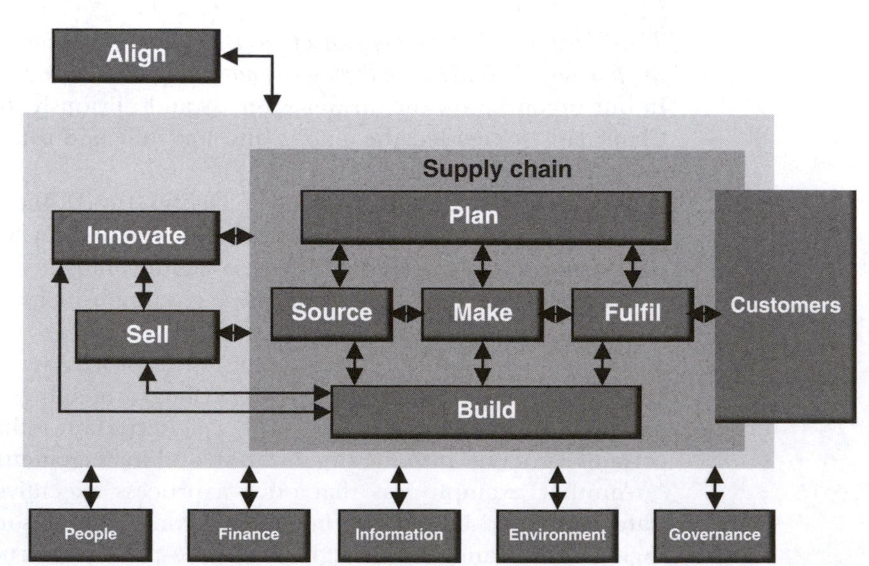

Figure 2.29 Enterprise process blueprint.

Firstly, it was the responsibility of the process owners to have well-established, aligned and collaborative relationships with the business units. The process owners must develop and share critical key performance indicators and targets, which must be continually updated and improved. SAP data was the source of nearly all metrics.

The global process management teams assess which specific performance measures and targets are applicable to the business unit. The key indicators of a process were seen as the leading indicators and the ability to predict performance. If a team's process measures indicate an increase in new customer signings, then it can predict revenue generation in the future. If signings are going down today, that predictive process measure alerts the team to take action now to prevent future revenue decrease. The corporate web-based dashboard mentioned earlier was developed so that process key performance indicator results can be communicated to each employee. All employees have access to the key indicators via the intranet. Employees can see how their team is going and provide feedback and make suggestions for improvement.

Given the geographical and business diversity of the organization, how did the organization go about standardizing its processes and resolve differences?

We used a four-pronged process management model that included: process leadership, design, performance and improvement, with SAP in the centre and we referred to this as our process model. Led by the SAP implementation team and the process owners, blueprinting sessions were held that ultimately led to 'process convergence'.

The goal was clearly to have one way of executing a process. Air Products did not complete a traditional 'As Is' process modeling activity, except at a high level, because we believed this would have slowed us down. The 'To Be' phase consisted of a review of the processes supplied by SAP as part of the system and we reviewed 'best practice' processes available from other sources and then determined a way forward. If there was any resolution necessary between business and function areas as to the proposed process it was resolved by the governance team.

After the first three or four years of process improvement, we still believed there were inefficiencies among the major processes, so we reviewed and began to organize our documentation and process linkages using the SCOR model and integrated it into our processes. This enabled us to move away from improvements in the individual processes and concentrate on where the processes intersect and where they link end-to-end. This is where we found the biggest areas for improvement.

Did you use an enterprise-wide process modeling tool?

We did purchase and install an enterprise process modeling tool, but have not deployed it broadly. We did use the mapping symbology conventions we established as part of the implementation. We mapped the processes in the businesses and then placed them in a central repository.

How did you embed and build upon this good process work within the businesses?

We established Communities of Practice (CoP) within the businesses or functions. This CoP comprised the Process Managers who met every month to share and explain success stories. This was a global network and over time, as the processes refined and deepened inside the business, the number of Process Managers increased from seven to eleven. Finally, a Supply Chain Director was assigned to each business unit with the accountability to employ and improve the standardized processes.

We also linked the BPM CoE with the Knowledge Management CoE to further embed both within the businesses.

As you progressively improved your processes what did you then do to maintain a level of momentum of improvement within the business?
One of the main things we found very effective was the creation of an intranet-based 'Idea Tracker' tool. This allowed everyone within the organization to input their ideas and allowed us to track them. We tracked these ideas up to the point at which they were either dismissed or became projects.

Every idea within Idea Tracker required the entry of the likely benefits to accrue from it. Over time, we required every approved, and yet to be approved, project to be entered by the business or function, to enable the tracking of company-wide project benefits. To support this, the company developed a rigorous definition of what constituted a benefit – was it ongoing, one-off, an avoided cost or reduced cost.

What have been the results of the 'Deliver the Difference' programme of work?
Well apart from the establishment of process governance structures and the shift in culture to being customer-based, during 2004 we doubled the rate of hard productivity savings and they tripled in 2005. We also created regional shared service centres and implemented the single instance of SAP across the organization. Customer Loyalty scores were 15% higher than prior to the start of the process/SAP implementation.

Key performance indicator	Impact
Selling, general and administrative (SG&A) expenses	−2.9%
Inventory (days on hand)	−2.4%
Operating return on net assets (return on capital)	From 9.5% in 2004 to 12.5% in 2007

What have been the lessons learned from your journey into being a process-focused organization?
There have been a number as you would expect. We would summarize them as:

- The most important decision we made related to process implementation. To implement the one instance of SAP was significant and the catalyst for change within the organization. Certainly it would have been more difficult without it.
- The Chairman's 'Deliver the Difference' guidelines added significant weight to this organizational change process. Prior to this the

organization was fragmented both geographically and culturally. The employees around the world, especially by continent, acted as independent companies. 'Deliver the Difference' was a trigger to galvanize the vision and the cultural change required by the organization.

- The resistance from middle management to a process-focused organization should never be underestimated and continues to be very strong. Middle managers feel the brunt of the change and there was an expectation that many of them would not have the skills to operate in the new world. We have learned to anticipate resistance to change and where this may occur. Change also takes a long time to occur, to be cemented within the organization, and convincing people to change is difficult. It is difficult to ask people to 'do the right thing' when they may not be in the same job. Training and reinforcement is necessary to make the desired behavior changes.
- Management must keep staff focused with the right priorities and only doing activities if it adds value to the organization.

We have also learned to:

- Focus on a few measurable initiatives
- Avoid creating a bureaucracy and overhead
- Define what we will and will not resource each year to keep process owners from being conflicted
- Integrate change plans into project plans and anticipate resistance from middle management
- Communicate frequently to leaders and then let them cascade and customize the message.

If another organization was to ask you for suggestions for its transformation towards a process-focused organization, what advice would you give them?
We would suggest:

- Do not do everything at once. Lay a foundation, and constantly build on what came before. We recognize and stifle the desire to jump to the answer too quickly.
- As you go step-by-step, sequence events with dedicated resources. *Transformation cannot be a part-time job.* Find a number of employees and isolate them to drive the initiative. Someone's job has to depend on this being successful. There needs to be no diversions or distractions. There is nothing like dividing the work, isolating a few steps, dedicating resources to them, creating credibility in the organization by meeting milestones and moving to the next step.
- Manage change aggressively. Anticipate resistance and deal with it early to gain acceptance that the change has value.
- Concentrate on visibility and velocity. By implementing a single instance of SAP, the company increased visibility of information. By increasing its focus on processes, it has achieved faster cycle times.
- Also concentrate on simplicity. Take complexity out of the business.

- Focus on both effectiveness and efficiency.
- Realize that the transformation process is never complete.
- Keep a customer focus. Do not have an internal focus for your processes. Ensure you have a metric(s) to focus on customer responsiveness during the transformation.

If you had to select just one lesson to share, what would it be?
This one is easy – *get CEO support. Moving to a process-focused organization requires too much change to accomplish without the strong endorsement of the CEO. Employees must see that the CEO is behind it and intends to make it happen.* When there is resistance to change, especially from middle managers, it needs to be escalated until it reaches the desk of the CEO. In his leadership sessions and town talks with employees all over the globe, John Jones often said: 'We have to face reality.' So the CEO must be prepared to say, 'You may not like it and you may think it is worse for your business, but this is what I believe we must do for the company to succeed.'

In Air Products' 2004 annual report, John Jones concluded with this paragraph:

> All of the groundwork we have laid since 2000 – restructuring our portfolio, resourcing our growth businesses and focusing on work process improvement – paid-off in fiscal 2004. Those four years presented great challenges for Air Products, but our focus and desire to win came through. And this will continue into fiscal 2005 and beyond. ...There's still much important work left to do. Our strategies will help drive top-line growth and return on capital. *And it will be our passion that delivers that difference.*

Part II

Management by Process framework overview

This part of the book is about the Management by Process framework that will allow an organization to become a process-focused organization and attain a high performance management environment. We provide an introduction to the framework by discussing organizations from both a functional and process viewpoint; an outline of the seven dimensions of the Management by Process framework; and then discuss each dimension in detail. During this detailed discussion, we will provide an explanation of why each dimension is important; what are key trends associated with it; what are the key elements; describe the visionary state (what should an organization aspire to); and finally we provide a roadmap of steps required to attain a sustainable process-focus and high performance environment. If an organization wishes to sustain its process improvement and management effort, then this is a roadmap that works.

Chapter 3

Introduction

If you really want to become process-focused, you are changing the way corporations have been evolving, which has been around functions, around business. Now what we are saying is that there is a new dimension called 'process', a whole new way of thinking about the way we add value.

(George Diehl, global director of the Business Process Center of Excellence and supply chain and process management education lead (retired), Air Products and Chemicals Inc., 2005).

In this introduction to the Management by Process framework – a roadmap to sustainable business process management – we will describe our vision for a truly process-focused and balanced organization. We recognize that this is quite distant from the current reality of most organizations so we will provide a roadmap of many of the steps that will be required to get there. However, it is important for any organization to know where its destination is, even though it may not reach it for some time. We will introduce the dimensions that we believe are essential within an organization if it is to achieve this visionary state.

So what is the current situation within organizations around the world? What are they doing now and how should they change if they wish to commence the journey to becoming more process-focused?

In this introduction chapter we will first provide a clear view on 'process' versus 'functional' view of the organization, and then present our Management by Process framework and introduce each of the seven dimensions that lead to a process-focused high performance management organization.

Functional versus process view

It is important to spend some time examining a functional versus process view within organizations; discuss how business processes become a topic to be addressed in the day-to-day management of an organization; and how an organization goes on the journey towards becoming more process-focused.

In order to start this journey the organization needs to know where it is going; what the end destination should be – *always start with the end in mind.* Ideally it should have agreement or at least have a general understanding of the current position. Once this is known the journey may begin.

But what does a process-focused organization mean?

Some of the current literature writes of a process-centric organization, which has been defined as 'an organization whose managers conceptualize it as a set of business processes' where they 'place their primary emphasis on maximizing the efficiency of processes, and not on maximizing the efficiency of departmental or functional units' (Harmon, 2003). While it is acknowledge that process-centric organizations will still have functional departments and divisions, they are seen only to exist in order to support the business processes and these functional units should be minimized as far as it is practicable.

We think 'process-centric' is an unfortunate set of words as semantically it implies that an organization needs to place its business 'processes' at the center of its thinking and acting. We would suggest that to do this *exclusively* would be just as inappropriate as the current situation where the total, or primary, focus is on the functional (silo) organization. To move from a totally function-based organization to a totally process-based organization may have us experiencing the 'same' but 'different' set of issues.

It is all about scale and balance, and the answer lies somewhere in the middle. What we need is an organization that is appropriately focused upon its business processes, and yet still capable of operating from a functionally based perspective to provide the performance, management and delivery of strategic objectives, all while maintaining a successful and profitable business and creating or maintaining the desired competitive advantage.

Figure 3.1 depicts how the process and functional views need to work together and support each other, while Figure 3.2 shows several of the responsibilities for each of the two areas.

While the authors are strong proponents of the need for organizations to place a higher priority on its business processes, the balance always needs to be kept in mind.

The journey towards being a process-focused organization will have a different starting point and path for each organization. The current misbalance between a functional management and business process-focus does however need to be addressed.

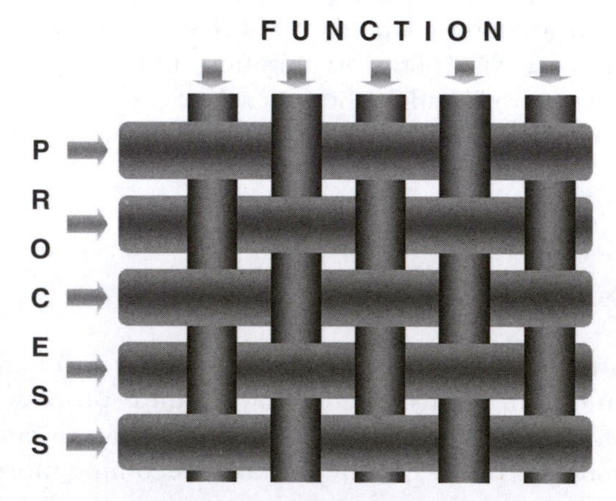

Figure 3.1 Tightly woven net of function and process.

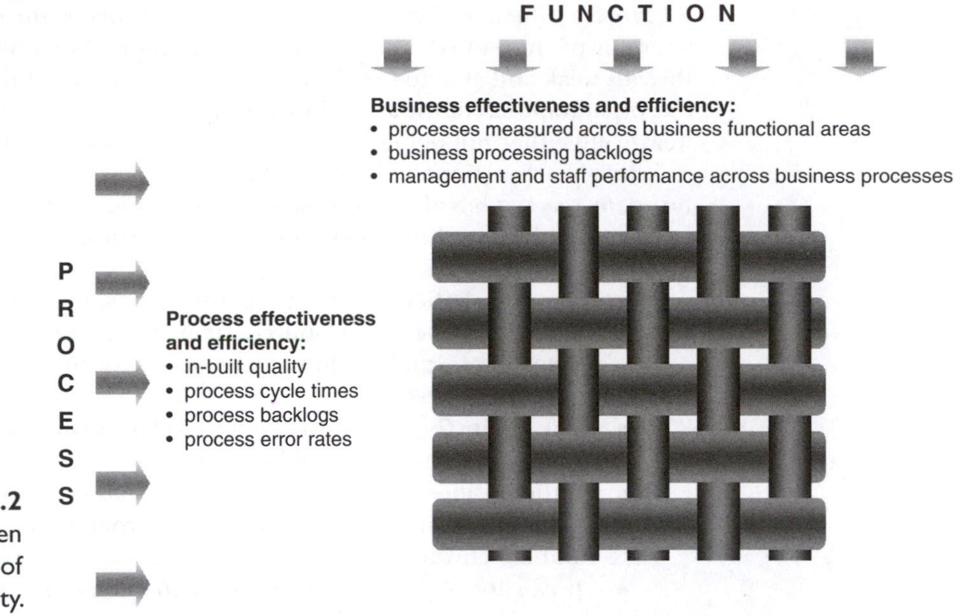

Figure 3.2 Tightly woven net and areas of responsibility.

Where an organization is heavily or even solely functionally orientated, then there may need to be an equally heavy weighting applied on the business process perspective, at least initially. The purpose of this initial 'heavy' process weighting is to counterbalance the current 'heavy' functional bias, both from a structural and thinking aspect. Over time, as the process-focus is adopted and accepted within an organization, both sides of the equation, functional and process, will need to be equally addressed as they merge into the one. An organization will know it has succeeded when it has one role fulfilling both a functional and process need, covering continuous process improvement and performance measure; and process management is 'just what we do around here'.

Currently most organizations are structured and managed with a functional bias, as shown in Figure 3.3. This figure shows how the typical organization has a totally functional bias and is paying no perceptible attention to its business processes (the organizational process view).

Where this is the situation within an organization, then the woven state of processes and functions are also unbalanced and weak, and look like

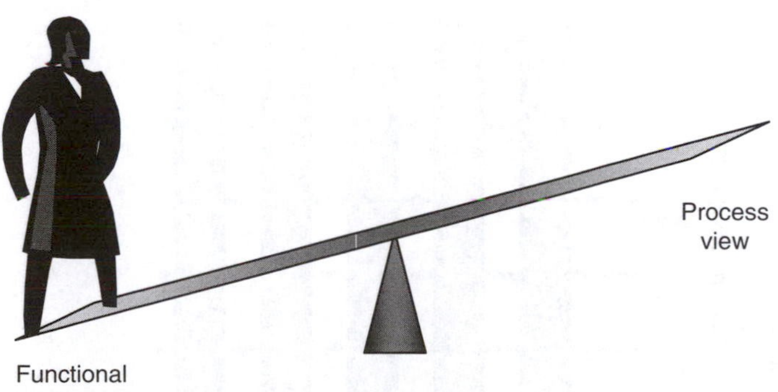

Figure 3.3 Functional bias in most organizations.

Figure 3.4 which shows that the weakened process threads threaten the strength of the entire fabric of the organization. Not only are the process threads weak and able to break, but there is also large 'white spaces' between the organizational fabric that allows 'things to slip between the gaps' in the weave and either not being managed in an appropriate manner or at all.

While the functionally biased organization has 'apparently' worked well for many years, it has allowed these process gaps to appear.

The advantages of functionally biased organizations are

- They are 'efficient for setting aspirations, making decisions, assigning tasks, allocating resources, managing people who cannot direct themselves, and holding people accountable' (The McKinsey Quarterly, 2007).
- The ability for an organization to focus a group of people, products and services into a profitable unit.
- If the business unit is large enough, it has enabled a complete business with all its processes, to operate what appears to be efficiently and effectively.
- It can focus the business unit on the achievement of some of the organizations strategic objectives.

The disadvantages are

- Lack of cooperation – 'the new element that can help 21st-century corporations create more wealth is large-scale collaboration, across the entire enterprise ... but to make people collaborate in large organizations, you must create a sense of mutual self-interest by holding talented, ambitious employees accountable not just or their own work but also for their performance in helping others within the organization' (The McKinsey Quarterly, 2007). Business processes are one of the means of providing the opportunity for this collaboration.
- The focus by one functional unit (silo) on the achievement of its results can be to the detriment of other parts/silos of the organization. This is especially true where a given business process cuts across functional boundaries.

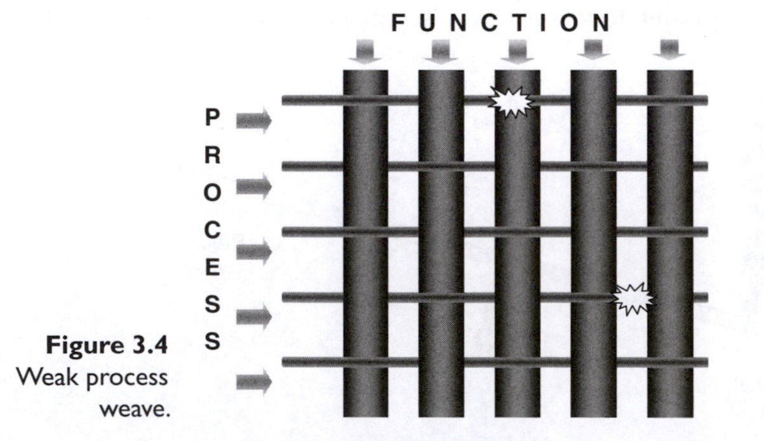

Figure 3.4 Weak process weave.

- Typically where business processes cross functional boundaries, there are often process inefficiencies. This can be particularly true where the functional boundaries are based upon an arbitrary grouping of activities. For example, hand-offs between an insurance underwriting department and the finance department.
- There may be investment funds available within one department that could be better spent on increasing the efficiency of another, and yet they are not released because of silo selfishness. As Spanyi (2003, p. 33–34) states: 'departmental silos and turf protection that impede performance are the natural enemies of process thinking and BPM'.

So how does an organization get from this weakened non-optimal current state towards a stronger organizational weave and ultimately the desired visionary state?

The first step is to provide management with an understanding of the benefits to be gained from being a process-focused organization. Acceptance of the need to move towards this type of organization is a critical step, and unless 'genuinely' agreed across senior management, then it will make the journey extremely difficult, and maybe impossible, to take.

Providing this knowledge and understanding, and gaining agreement, is predominantly the role of executive leadership – not just from an inspirational perspective, nor creating and inspiring management and staff, nor just walking the talk (behavioral). Executive leadership's role is to create the structures, incentives (rewards) and individual targets (KPIs) to drive towards the process-focused goal. Get the KPIs wrong and it is all at risk – possibly terminal risk.

If executive buy-in cannot be gained in the initial stages of the journey, it may be necessary to 'prove' the value of process improvement and process thinking to management as well as staff within the organization. Depending on the organization this may occur through a manager in a 'pocket' of the organization starting an 'under the radar' initiative that may have the opportunity of progressively convincing the entire organization of the importance of the process view.

Case study: Lack of organizational strategic alignment

We know of one large government instrumentality who recognized the inadequacies of the functional (silo) view in 1997 and has been trying to convince the functional executives to 'get on board' the process view ever since. Significant progress has been made in the last year or two with a specific project to progress towards agreement on the appointment of process stewards (for end-to-end processes across functional boundaries); a central investment strategy based largely on its business process needs; and aligning its significant investment funds with the organization strategic objectives. This last point may seem like an obvious thing to do, but an internal analysis of its projects revealed that 87% of projects currently being executed were not aligned to the organizational strategy.

Message: If your projects are not adding value to the organizations strategic objectives, then why are you doing them?

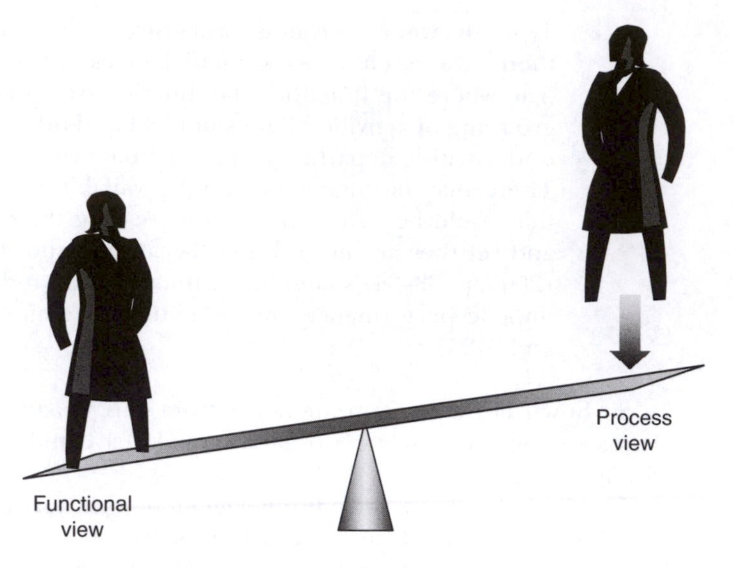

Figure 3.5 First steps on the journey.

Certainly of recent times, more and more organizations have recognized the problems with a solely functionally biased structure and started to understand that its business processes can contribute significantly to the performance of the business and hence have started to pay more attention to them. There is a growing understanding that managing the linkages between processes across the organization is of equal importance to managing the cross-functional relationships. Figure 3.5 shows this process view starting to occur.

The intertwined nature of the process and functional views is an interesting one and begs the question, 'how far should the right-hand side of the lever (process view) be depressed downwards?' The strictly process-centric view may suggest that it should be depressed downwards as far as it will go (or at least to a significant depth) and perhaps until the functional view disappears entirely. That is, an organization has totally transformed from being functionally bases to process bases.

It is argued that changing from the traditional functional, hierarchical orientation to a process-focused orientation will mean that organizations will function with greater efficiency and effectiveness, to the benefit of management, staff, customers and all other stakeholders.

If we explore this reasoning for a moment, then the organization may look like Figure 3.6.

While Figure 3.6 shows how both parts of the organization (process and functional) are intertwined (which is a good thing), it also shows that while the horizontal process lines have thickened and therefore become stronger within the organization, the vertical functional lines have become thinner and less important with some breaking. Thus, the functional view has become the weak link or component within the organization and, in fact, some are starting to 'break' or not work very well. The impact of this in practice could be:

- A potential weakness of a 'complete' process view is that as we focus on all the *individual* business processes to make them optimal, we have assumed that the *sum of the individually optimized processes will be better for the organization than the previous functional view.* Even if the

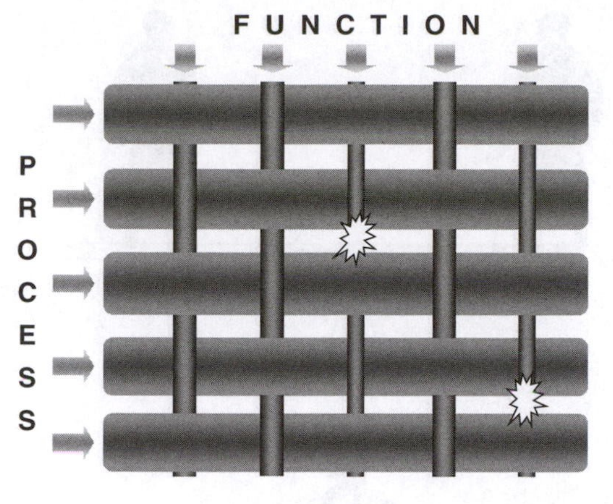

Figure 3.6 Weak functional weave.

total of the optimal processes actually is greater than the previous functional view, it may not be as optimal as it could be.

- While the *individual* end-to-end business processes themselves appear to be functioning efficiently and effectively, the organizations resources may not be optimally allocated across the various business processes. Resources (people and capital investments) need to be balanced across all the organizations key business processes in order to create the very best optimal situation and it is often the addition of the functional view that achieves this.
- If the process view was taken to its extreme, each business process area could have their own IT department and staff who would maintain the supporting technology for that particular process. So if an organization had eight to ten major processes, IT would be split across these eight or ten areas. This has the potential to create: conflicts (as each group tries to simultaneously enhance or change the same business application); duplication and additional expense for the organization. This is clearly not in the interests of the entire organization and will need to be managed from a functional view, working with the end-to-end process view. The same is true of other business support areas, such as, Human Resources and Finance.
- Just as functional managers have developed their own silo's and 'power bases', what is to say the process managers (process owners/ stewards) will not do the same?

It could be viewed that the initial stages of a changing organizational structure that is progressing towards a process-focus is a compromise between the process and functional views. But compromises always contain trade-offs from the ideal situation and this may in fact be the case in these initial stages. However, as both the function and process view of an organization are important, getting the balance correct is of course the challenge, as depicted in Figure 3.7, and knowing when to stop pushing down on each side is difficult.

We have seen organizations who have attempted to create the balanced position in one restructure. This meant that the traditional functional managers were 'suddenly' faced with the additional responsibility of managing the end-to-end business processes as well, without a thorough understanding or

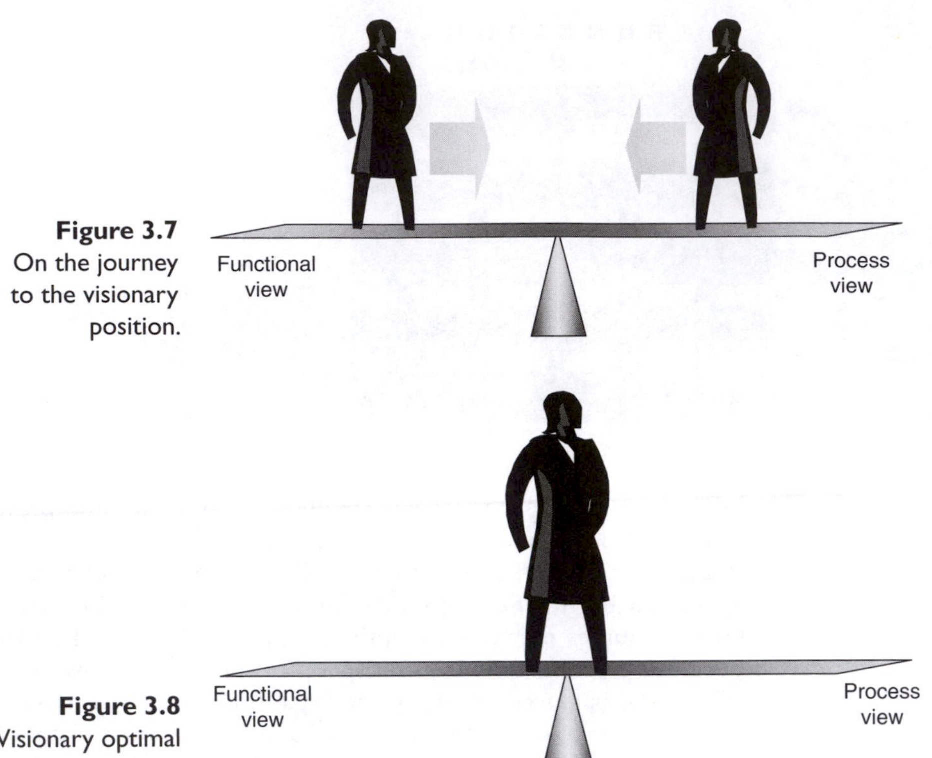

Figure 3.7 On the journey to the visionary position.

Figure 3.8 Visionary optimal position.

appreciation of the role and responsibilities that accompany it. This may prove too ambitious for most organizations in the early stages and too much for a line manager to 'juggle two hats' when he has no experience with the process view.

As part of this transitionary journey, it may be more appropriate for the organization to have different people completing these roles, the functional or line managers and others representing the process view. An example of this is the process steward role that we will discuss in detail in Chapter 5. This role may be completed by people who have no other responsibilities. These initial process roles will cut across the organization, impacting or effecting several functional silos within the one organization, especially if, as it must, cover a process from an end-to-end perspective.

When the organization is ready and has grown in its process maturity, it will progressively see both sides moving towards the middle so that the organization's management will no longer be separated into distinct functional- and process-based roles. As the organization gains deeper experience and appreciation for both the process and the functional view, this will enable a joint and more balanced management who will span both views simultaneously, as shown in Figure 3.8 where functional and process responsibilities will just become 'what we do around here' and part of a managers responsibilities.

As managers, or indeed staff, complete their tasks, they will always be considering the process implications. The people within the organization will have a culture that supports and encourages a continuous process improvement thought process and the capability and skills to make it a reality. The number of separate improvement projects will reduce, as improving the business processes will be a continuous activity. Managers will be working together

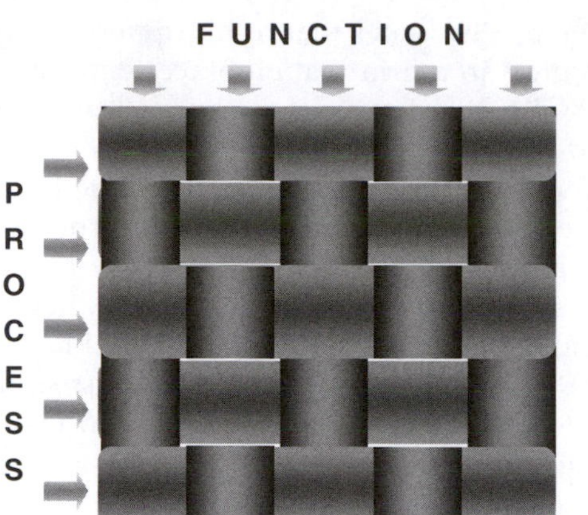

Figure 3.9 Function and process weave of visionary state.

in a collaborative way, where an end-to-end process spans across several functional areas and thus manages. This requires considerable communication and collaboration, as the process and functional needs must be balanced. Managers and team leaders will be using the process and people performance reporting measures that are produced to manage their contribution to the business. This balanced state is depicted in Figure 3.9.

Always remember that:

> Creating a process-focused organization must be in the context of something that is very important to the company. A process-focused organization for the sake of being process-focused has little value and, quite frankly, has a negative feel to it. If it is not placed within the context of the strategy and direction of the company, you are probably not going to get a whole lot of recognition or buy-in for the concept itself.
>
> (Bill Cantwell, vice president of supply chain, executive process owner 'Sell', and chair of the global process board, Air Products and Chemicals Inc., 2005).

We have seen many organizations where functional managers consider their 'silo' as their own 'territory' or 'fiefdom', whereas in the balanced organization, all managers would have the benefit of the organization in mind rather than their 'territory'. In the woven net of functions and processes, the whole, rather than the individual parts, must be of primary focus. There will usually be several stages in a journey to progress the maturity of the individuals and the organization as a whole. The behavior of individuals and especially executives and managers within an organization is not only a reflection of the organizations culture, but significantly influenced (in fact, often driven) by the targets (KPIs) set by executive leadership. It is an interesting phenomenon that while executive leadership understands that targets (and especially financial rewards) drive executives and management behavior, and yet there is a significant reluctance within most organizations to change the targets and reward systems to support the required behavior to support process-focused performance. We will discuss this in more detail in Chapter 6 as it is an important aspect of any organization, and critical component in the drive towards a balanced process-focused organization.

Is this transition to a process-focused and balanced organization easy to achieve? Definitely not!

Spanyi (2003, p. 139) states that in his opinion, 'anywhere from a quarter to half of executives in a firm that embraces enterprise management *(a process-focused view)* will either leave or be asked to leave within a year of the effort being launched'. So the 'super highway' to the process-focused visionary state is not a smooth one, but the rewards once achieved (as shown in the previous case studies) are worth the journey for all concerned.

Within the complexities that exist within the corporate environment the journey is not straightforward nor easy and sometimes extremely difficult to see or envision. There will certainly be road blocks, dead ends and a few U-turns along the way, but an organization must start somewhere and continue to strive to complete the journey if it is to maintain or obtain a competitive advantage in the market.

In order to enable organizations to achieve a process-focused state, we have simplified this journey as much as is possible, we have broken the vision and pathway towards it into seven super highways (dimensions) that may be traveled largely simultaneously.

Overview of Management by Process framework

If an organization wishes to take a leap towards becoming more competitive within its particular marketplace, this process-focused journey is one of the important areas it will need to address. We will now build an understanding of our Management by Process framework.

Strategy

The foundation of our Management by Process framework starts with the formulation of an organizations strategic objectives and ends with their fulfillment (Figure 3.10). Unfortunately some managers appear to stop after the creation of the strategy or objectives.

Strategy is the area on which many executive managers spend a considerable amount of time, effort and resource. This dimension is where executive management, general line management and staff all need to become process aware, fully understanding the power and difference that changes in business processes can make within an organization. The best way to start is for the executive management to gain a clear understanding of that business processes can and will make a significant difference to the performance of their organization and then establish plans to move towards process-focused high performance management. They should back this up, not just with encouragement and instructions, but also setting their own and line managers KPIs to reflect this process-focus.

Execution matters

However, the strategy is only one part of the journey towards success and could be considered the easier part. Formulating a new strategy seems to be

Figure 3.10 Management by Process framework: strategy.

easy compared to its execution especially from a green field situation where you do not have to face the grim reality and trade-offs of day-to-day operations. We believe that execution is the challenging aspect for most organizations.

This is supported by two articles in the Harvard Business Review (January, 2008) where Prof. Cynthia A. Montgomery stated:

> What we have lost sight of is that strategy is not just a plan, not just an idea; it is a way of life for a company. Strategy doesn't just position a firm in its external landscape; it defines what a firm will be. Watching over strategy day in and day out is not only a CEO's greatest opportunity to outwit the competition; it is also his or her greatest opportunity to shape the firm itself.
>
> (Montgomery, January, 2008, p. 56)

Kaplan and Norton also support the view that execution is critical when they state that:

> breakdowns in a company's management system, not managers' lack of ability or effort, are what cause a company's underperformance. By *management system*, we're referring to the integrated set of processes and tools that a company uses to develop its strategy, translate it into operational actions, and monitor and improve the effectiveness of both. The failure to balance the tensions between strategy and operations is pervasive: Various studies done in the past 25 years indicate that 60% to 80% of companies fall short of the success predicted from their new strategies.
>
> (Kaplan and Norton, January, 2008, p. 64)

> A business leader's most important job is the execution of plans, the 'detail work', making sure that the staff is getting results. This is the sort of responsibility that cannot be delegated. It is the leader's primary duty to see that every member of the team is carrying out his part of the big plan to ensure the whole company's success. There are no excuses for failure: the market will always be tough. What spells the difference between successes and failures is the ability to execute plans.
>
> (Bossidy *et al.*, 2002)

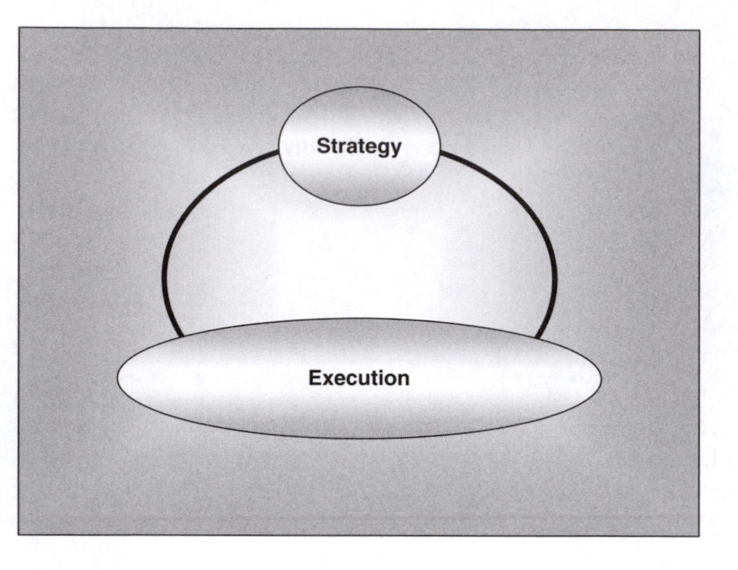

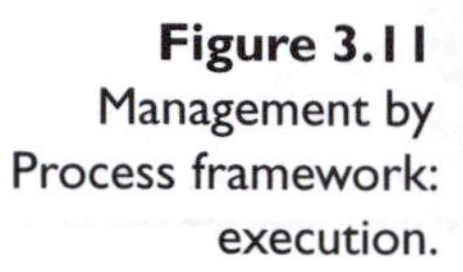

Figure 3.11 Management by Process framework: execution.

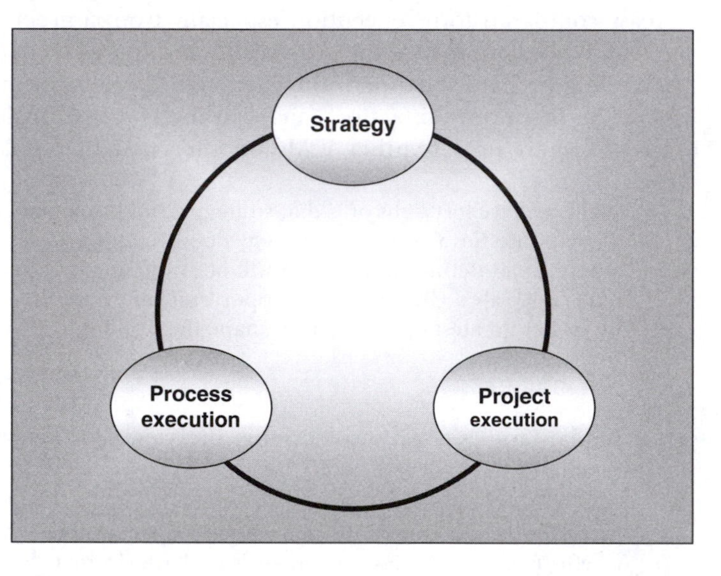

Figure 3.12 Management by Process framework: process and project execution.

Hence, execution is the next component (Figure 3.11).

However, we believe it is essential to distinguish between two types of execution (Figure 3.12):

1 Project execution
2 Process execution.

Much of the process management and business improvement literature focuses only on one of these elements, resulting in an incomplete picture on how to be successful within an organization.

Project execution deals with the execution of projects, programmes or portfolios that are distinct unique activities that contribute to well-defined predefined deliverables with identified resources and a definite start and end date. Projects, programme and portfolios must always be geared towards

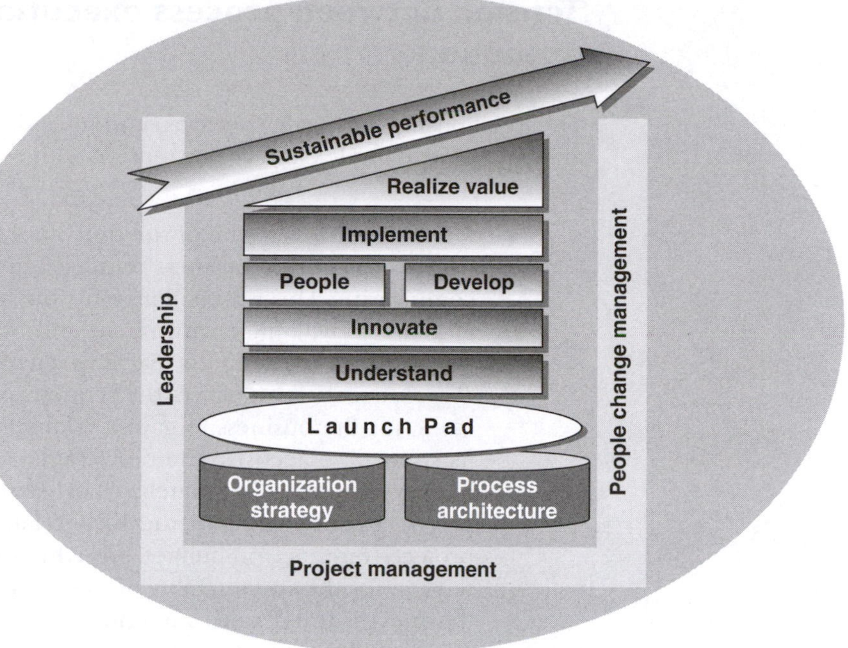

Figure 3.13 Management by Process framework: 7FE Project Framework (project execution).

handing over their work to the business to ensure sustainability. (Although we recognize that a few businesses actually function as project execution organizations – it is all they do, for example, engineering businesses. In this instance, the normal business as usual projects are process execution, while projects to innovate and change are labelled as projects.)

Project execution is the area where the organization must establish a sustainable business process improvement programme and project structure that not only suits the particular organization, but also enables timely, successful, repeatable projects, delivering real benefits to the organization. Benefits that are *actually* delivered and *known* to be delivered, not just noted in the business case in order to obtain project approval, and 'assumed' that they have been or will be delivered. This can only be achieved with the provision of an appropriate level of funding, resources and a process-based project management framework to complete the work. How an organization may go about the successful completion of this component has been discussed in Jeston and Nelis (2008) (Figure 3.13) and will be examined from a project management methodology and implementation perspective in Chapter 9.

Process execution deals with the operational running of the business processes, allocation of people and use of technology – it is often called operational management or 'business as usual'.

Process execution requires a pro-active, dedicated and passionate approach. We have called this High Performance Management which relates to the management and utilization of process, people and technology, and much more. To be effective with High Performance Management all of the seven dimensions highlighted in this book need to be in place and executed to a high standard (these will be described a little later in this chapter).

Tension between process execution and project execution

The relationship between *project* execution and *process* execution is simple and yet can be complex and create tension within an organization.

- *Process* execution – one of the outcomes from this activity (the business) is to provide business requirements for projects to build and comply with. This is a challenge for the business because *process* execution is a dynamic environment and by the time the *project* execution delivers, the business may have changed.
- *Project* execution – delivers the requests of *process* execution. In most cases, for the business to gain a competitive advantage via meeting its strategic objectives, the deliverables will need to provide a step improvement. Unfortunately, when setting the success criteria for project execution, often the KPIs relate to the completion of the project (on-time, on-budget, with high quality), which can create tension for the sustainability of the solution being provided into the 'business as usual' environment.
- *Process* and *Project* execution share the same people and technology and impact largely the same processes. Most of the tension in an organization results from conflicting priorities and views on the usage of resources (processes, people and technology). These tensions do not just occur at an organizational level, but also occur with most managers and staff (subject matter experts) as they too must divide their time between projects and process execution.

Process, people and technology

Execution cannot take place without these three components – they are essential to the achievement of results in both areas. Process execution (operational business as usual) cannot occur, to a high standard, without people, supported by technology and being able to understand business process performance. There needs to be an appropriate equilibrium between project execution and process execution and these three key dimensions (Figure 3.14.

We do not specifically discuss process execution in its own chapter as we see it as the culmination, or result, of all seven dimensions being executed well.

Process governance

Process governance is required to ensure that the strategy, project execution and process execution perform well *and* are aligned. Many organizations have a fragmented approach towards process governance – there may be some level of governance on projects and, little to none on business processes. However, there is rarely governance on strategy and the interface between strategy, project execution and process execution.

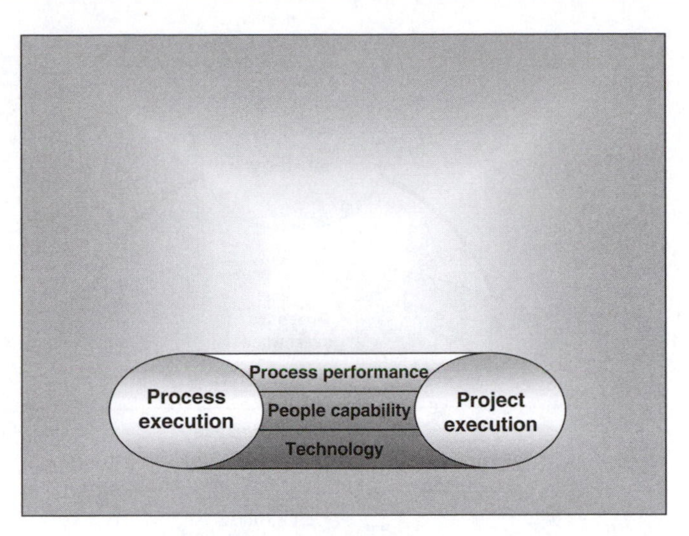

Figure 3.14 Management by Process framework: process, people and technology.

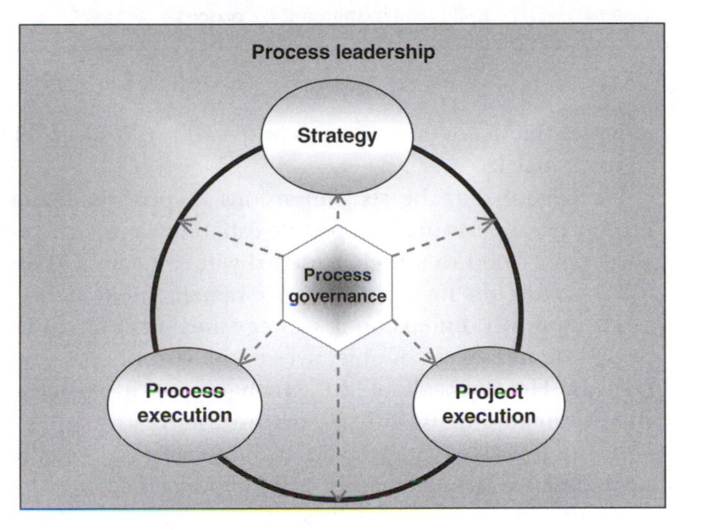

Figure 3.15 Management by Process framework: process governance.

It is important to understand that governance will have different aspects (roles) for the governance of strategy, the governance of process execution and the governance of project execution. Different governance will also be necessary for the links between strategy and process execution; strategy and project execution; and process execution and project execution. These interfaces must ensure that all dimensions are systematically and continuously linked, as shown in Figure 3.15.

Bringing the dimensions together

Figure 3.16 shows an overview of all the dimensions for an ideal process-focused management framework and how each component interacts with the business and each other.

We call this the Management by Process framework and it refers to the normal day-to-day functioning of an organization – the day-to-day operational

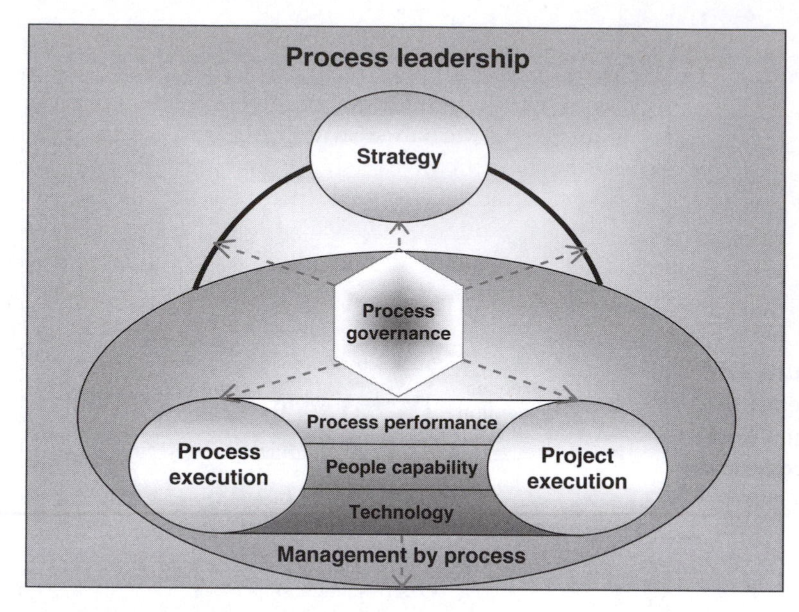

Figure 3.16 Management by Process framework.

activities that leaders, managers and staff are responsible for performing to a high standard.

We believe that the six dimensions of process leadership, process governance, process performance, people capability, project execution and technology when performed to a high standard will result in successful process execution and therefore the high performance management an organization via process.

The seventh dimension is strategy and it is critical in providing the necessary guidance – a set of objectives and a vision – for the organization to strive towards. However, all of the Management by Process framework activities must be in alignment with the organizations strategy.

In the following chapters we will describe each of the seven dimensions in more detail and each chapter will include:

- a description of why the dimension is important
- some key trends
- the key elements
- a brief description of the ideal or visionary state
- a road map of the journey for the organization towards the visionary position.

We have provided below a brief overview of each dimension.

Process leadership, as the description implies, only covers leadership from a *business process* view point. We will discuss the differences between leadership and management by examining a *transformational* and *transactional* leadership model. It is the leaders within an organization who are responsible for driving its purpose and igniting the passion to build and create the organization's sustainability in a globalized world. This kind of sustainability builds when you have total alignment with the key aspects of the financial position, the people, business processes and organizational strategy.

Process governance is arguably one of the most important dimensions for the continued and long term (sustainable) success in creating a process-focused organization. Without a level of leadership commitment to the establishment of a governance structure based around business processes, it will be extremely difficult to achieve anything except isolated business process improvement projects and activities. These isolated business process improvement projects and activities may indeed be the first steps for a process immature organization. In this chapter we examine the various roles and their respective responsibilities; how to establish and manage a business process performance–based environment, with commensurate reward systems; and the process management controls that are necessary.

Process performance, that is, the measurement of business processes, is a significant part of the sustainability of process improvement and process management within an organization. If you link process performance to the accountability and responsibility for business processes (in process governance) then you have a powerful ability to actually manage your business effectively and efficiently in a sustainable manner. This chapter is about understanding what process performance is; how it should be applied within an organization; and how it must be linked to the other dimensions to achieve an organization that is managed by process and reaches a high performance management position.

Strategic alignment is about ensuring that business processes are aligned with the strategy and objectives of the organization. Without this alignment the business processes will eventually deviate from the key drivers and goals of the organization, further alienating business process management and improvement from the business process execution and management. This will lead to a sub-optimal competitive advantage for the organization.

People capability is the center piece via which a process-focused organization is created and sustained. In this chapter we describe how to build internal capability within an organization, why it is critical and must be supported by the creation of a Center of Business (Process) Innovation (CBI) if the journey towards a visionary state is to be achieved. These aspects will only work effectively if the engagement model between the business, the center of business innovation and information technology is appropriate and works.

Project execution – few people would argue that for an organization to have a strategy is a good thing. Few people would also argue that having a strategy is not very helpful to an organization unless it is able to be executed and executed well. Yet, organizations generally find executing strategy *effectively* a very difficult activity. Projects are the primary vehicle for implementing strategic changes. If the senior management team either chooses the wrong projects, the wrong scope or implements too slowly, the organization fails to meet its goals. In this chapter we will examine the critical activities to executing projects significantly more effectively than is currently the case in most organization.

Technology – while we are big fans of technology, there are already many books which have covered this in more detail than we have time for in this book, so we have covered technology in Appendix D and have only covered it to provide a brief explanation of the latest aspects and endeavored to explain it in simple terms so that non-technical people can understand it.

Before we take each of the dimensions and examine them in the following chapters, we would like to complete the discussion we started in Chapter 1 with regard to the Kaplan and Norton Management System model and the analysis of de Waal characteristics of a High Performance Organization.

Management System model

In Table 3.1 we have provided an analysis of the Kaplan and Norton (January, 2008) Management System model and our Management by Process framework. In the first column is the Kaplan and Norton stages and main tools and activities, we have then provided a link to our framework Dimensions, together with some commentary.

We believe that the Management by Process framework links within this model in relation to execution and strategic alignment and provides additional tools to ensure that the strategy, via projects and processes, is executed successfully. However, no framework or approach is perfect or sufficient in

Table 3.1
Kaplan and Norton Management System model compared to Management by Process framework

Kaplan and Norton Stage and main activities	Management by Process framework dimensions	Comments
1 *Develop the Strategy*: • Formulate mission, vision and values • Strategic analysis • Strategic formulations	Strategic Alignment	Strategy formulation is not in the scope of our framework, as there are many excellent books around on this topic. However, we see the high level process analysis as an important part of strategic analysis to determine the key processes that provide competitive advantage. Leadership plays a crucial role in providing a vision.
2 *Translate the Strategy*: • Produce Strategy Map • Define Strategic Themes • Develop Balanced Scorecard of Performance Metrics • Identifying and authorizing resources	Strategic Alignment Process Performance	Translating Strategy to Execution is critical. We have provided steps of how to do this and how the strategic decision will impact the processes and projects. We suggest that the Balanced Scorecard includes a reference to key processes to ensure a seamless hand-over to business as usual (operations) at the end of a project as well as ensuring alignment of strategic objectives and process targets.

(Continued)

Table 3.1 (*Continued*)

Kaplan and Norton Stage and main activities	Management by Process framework dimensions	Comments
3 *Plan Operations*: • Initiate process improvements • Prepare detailed sales plan • Prepare Resource capacity plan • Specify dynamic operation and capital budgets	Project Execution Process Execution People Capability	We have split operations into two components: the project and process execution (business operations). The Launch Pad Phase of our 7FE improvement framework (Jeston and Nelis, 2008) ensures that improvement initiatives are focused on achieving the "right" objectives and value for money. It is important upfront to specify the rewards and remuneration of all people concerned.
4 *Monitor and Learn*: • Operational review meetings • Strategy review meetings	Process Performance Project Execution Process Execution Process Governance	Monitoring the processes and projects is vital to ensure that they are executed well and provide the required outcomes. However, monitoring is just one aspect of managing these processes and projects. Clear process roles and responsibilities, including ownership, is essential. Continuous process improvement will never be successful without either managing the processes and the improvement projects.
5 *Test and Adapt the Strategy*: • Costs and profitability report • Statistical analyses • Emergent strategies	Process Performance Strategic Alignment	Testing the strategy is inherent to our Management by Process framework, especially through Process Performance and the Process Governance of projects, programs and portfolios. While adapting the strategy is outside the scope of our framework, ensuring strategic alignment is critical.

itself. It needs to be adjusted for the specific organization and the leaders and people have to believe in the journey and strive to achieve excellent results, hence the emphasis we place on Process Leadership.

Characteristics of a High Performance Organization

While the analysis of De Waal (2005) derived 68 characteristics of a High Performance Organization (HPO), we have chosen to review the top 20 because we believe these are the most critical requirements for the HPO.

Table 3.2 shows these top 20 grouped according to the 8 factors. Again we have linked them to the dimensions in the Management by Process framework and provided comments. In addition, we have shown the weighting factor (as a percentage) as derived by de Waal.

Table 3.2
Top 20 characteristics of High Performance Organizations

De Waal Top 20 HPO Characteristics (weighting factor – %)	Jeston and Nelis Management by Process dimension	Comment
External		
Continuously strive to enhance customer value (61.1%)	Overall Strategy Alignment Process execution Project Execution	The need for Customer-focused solutions is highlighted throughout this book. The Strategic Alignment focuses on the way the processes provide value to the customer. Process and Project Execution are customer-focused.
Maintain good and long-term relationships with all stakeholders (38.3%)	Process Leadership	The leader must to provide internal and external guidance and facilitate key stakeholders.
Monitor the environment constantly and respond adequately (31.2%)	Strategy Alignment	Strategies needs to be consistently monitored and are updated on an increasingly rapid scale. Agile strategy processes are critical to ensure swift results.
Choose to compete and compare with the best in the market place (26.2%)	Strategy Alignment	Strategy will set the objectives for the organization. There is no room for complacency or lack of ambition.
Organization design		
Stimulate cross-functional and cross-organizational collaboration (29%)	Process Performance Process Execution	We suggest end-to-end processes that normally will cross-functional and organizational boundaries. Often the human touchpoints between functions and organizations are critical bottlenecks. Collaboration is the start to address these problems.
Simplify and flatten the organization by reducing boundaries and barriers between and around units (28.3%)	Process Performance Process Execution	Processes are increasing becoming networks that need to be seamless to ensure competitiveness, value to the customers and profits to the participants.

(Continued)

Table 3.2 (*Continued*)

De Waal Top 20 HPO Characteristics (weighting factor – %)	Jeston and Nelis Management by Process dimension	Comment
Strategy		
Define a strong vision that excites and challenges (24.6%)	Strategy Alignment Process Leadership	Leadership must ensure that the vision of the organization is something that employees and partners can associate with and are eager to strive towards its realization.
Process		
Design a good and fair reward and incentive structure (49.2%)	Process Performance People Capability	We recognize that most people do what they are rewarded for. This is neglected or ignored too often.
Continuously simplify and improve all the organization's processes (45.8%)	Process Execution Project execution Process Performance	Process improvement can be initiated as a project or positioned as part on continuous process improvement. It is important that both these type of improvements follow a structured approach to their realization and are embedded in management by process.
Measure what matters (35.8%)	Process Execution Process Performance	*"You can't manage what you don't measure".*
Report to everyone the financial and non-financial information needed to drive improvement (32.1%)	Process Execution Process Performance	Process execution contributes to the realization of strategic objectives. We suggest leveraging this link when improvement opportunities are identified or improvements activities require escalation.
Continuously innovate products, processes and services (31.5%)	Process Execution	Management by Process framework incorporates that processes are continuously managed, monitored and improved
Technology		
	Technology	None of the technology characteristics have made it to the top 20. Underlining our firm believe that technology can provide significant benefits, but by itself is insufficient.
Leadership		
Maintain and strengthen trust relationships with people on all levels (46.1%)	Process Leadership	Process execution and process improvement require the transcendence beyond the "mechanics". It require believe, spirit and commitment. The case studies in Chapter 2 show the importance of inspirational leadership.

(*Continued*)

Table 3.2 *(Continued)*

De Waal Top 20 HPO Characteristics (weighting factor – %)	Jeston and Nelis Management by Process dimension	Comment
Live with integrity and lead by example (40.5%)	Process Leadership	"Walk the talk".
Apply decisive action-focused decision-making (26.5%)	Process Leadership	Leaders are people who are willing to make clear-cut decision, though sometimes controversial and they encourage other employees to do the same.
Coach and facilitate (23.7%)	Process Leadership People Capability	Leaders need to coach and facilitate their staff and not overwhelm them.
		External coaching on new topics, such as Management by Process and Business Process Management, can bring momentum and a great deal of experience.
Individual and roles		
Create a learning organization (51.4%)	Process Execution	We suggest System Thinking where lessons learned is structurally and systematically embedded in the "*way we do things around here*".
Attract exceptional people with a can-do attitude who fit the culture (31.8%)	People Capability	Process Management requires a new way of thinking, a result-oriented attitude and in our experience learning new skills is easier than changing attitudes, hence the need to attract people with the right attitude.
Culture		
Empower people and give them freedom to decide and act (56.7%)	Process Leadership People Capability	We clearly state that good leaders empower their staff and allow them to make their own decisions. Loners and dictators are not good leaders for organizations.
Establish strong and meaningful core values (25.2%)	Process Leadership People Capability	Business Process Management needs to live in the hearts and mind of all people concerned. Formulating these values is easy; the challenge is to establish them.

Chapter 4

Process leadership

Written with Mandy Holloway

Introduction

This book is about the management of an organization from a business process-focused perspective. While this chapter is about leadership, we will only review this huge topic from a *business process* view point (Figure 4.1). We will discuss the differenc es between 'leadership' and 'management' by examining a *transformational* and *transactional* leadership model.

The real issue for leadership within organizations is how are they going to respond to the greatest challenge since the Industrial Revolution – globalization. To put this in perspective did you know that (Fingar, 2007):

- The 25% of the population in China with the highest IQs is greater than the total population of North America.
- If you are one in a million, in China there are 1,300 people just like you.
- The country with the most English-speaking people is not the United States of America. It is China.
- In the 1980s, capitalism supposedly triumphed over communism. By the year 2050, communist China is expected to have a gross domestic product (GDP) twice that of the United States of America.
- It is estimated that 1.5 exabytes (1.5×10^{18}) of unique, new information was generated worldwide in 2007. That is estimated to be more than in the previous 5,000 years.
- The amount of new technical information is doubling every two years. It is predicted to double every 72 hours by 2010.
- The US Department of Labor estimates that today's learner will have 10–14 jobs…by age 38.

As Fingar says, 'shift happens'. And it is happening faster all the time. So how does organizational leadership cope with this?

It is the leaders within an organization who are responsible for driving its purpose and igniting the passion to build and create the organization's

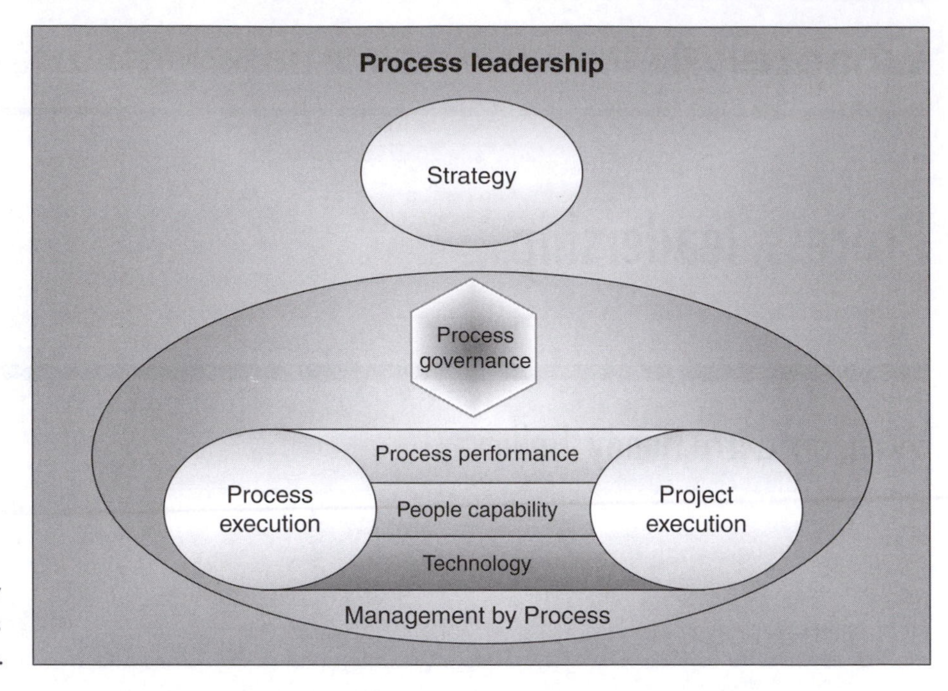

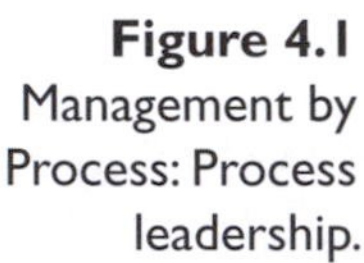
Figure 4.1 Management by Process: Process leadership.

sustainability in a globalized world. This kind of sustainability builds when you have total alignment with the key aspects of the organizational strategy, business processes, the people and the financial position.

It is the leaders in an organization who need the myopic focus and energetic passion to live this alignment everyday and create the infectious commitment that is needed to keep everyone on board within the organization!

> Leaders need to set out the expectations and ensure everyone's needs are taken into account within the promises they make – leadership builds trust and a 'can-do' attitude within the internal and external stakeholders of an organization everyday.

To build this kind of sustainability, to get it right, to create this infectious commitment many organizations are investing substantially in time, money and emotions to identify their key business processes and build an environment to support them everyday. They are beginning to see key business processes and an organizational culture of continuous business process improvement as one of the key foundations of operational sustainability in their organization. Conceptually, many leaders are beginning to understand that key business processes are able to provide a competitive advantage in the marketplace. They are beginning to understand that a process-focused culture to surround and support this is something that everyone needs to commit to, build and maintain for the organization to sustain competitive advantage. The organization will need to:

- attract customers, generate revenue and produce strong bottom line financial results
- be an attractive place to work and belong
- be a highly respected and trusted brand
- be a desirable and trustworthy partner
- add value to the national, global economy and contribute to society.

Case study: Alignment of organizational messages and marketing

We are constantly amazed at the lack of alignment within organizations. Recently one of the authors sat in the foyer of a global business and watched in awe as an amazing message featured on the huge plasma screen attached to the wall. The message left us inspired and excited about what this business stood for. So you can imagine the shock we felt when greeted by the person we were visiting – a person who was employed as a senior executive in the business – when they deigned ignorance about the message as we went up in the lift and fobbed it off as another one of those marketing gimmicks that everyone forgets to tell us about! How much money had the organization spent on the marketing campaign and we can only imagine the damage it may cause in failing to deliver on brand promise.

Message: Leaders need to 'walk the talk' *all* the time and build infectious commitment for the future direction. They are responsible for creating alignment and preventing this lack of cohesion.

Having said this, leadership is a complex and multi-faceted activity. We were very interested to read on 4 September 2007 the Australian Financial Review article (p. 58) by Alistair Mant (who is the chairman of the UK-based Socio-technical Strategy Group and an author) as it substantiates many of our thoughts on leadership. It is worthwhile spending a few moments reviewing this article as it comments clearly on leadership and management.

Mant stated:

> the world is full of evidence that we can't run things very well. We have become obsessed by 'leadership'. The less we like to trust our leaders, the more we overuse the L-word. We can't be sure what it means any more.
>
> Meanwhile, the M-word – management – has been relegated to also-ran status. Yet we are surrounded on all sides by cock-ups, blunders and other evidence that we can't run things properly.
>
> If 'managing' doesn't mean running things properly, what does it mean?
>
> Overwhelmingly, when the employees of large organizations are surveyed about their de-motivators at work, the most common complaints concern perceived incompetence of superiors and what is generally known as 'office politics'. In other words, the people who really matter in the interface with customers, clients and the marketplace know their bosses are often incompetent and self-serving.
>
> It is a matter of judgment whether we describe these as failures of 'leadership' or failures of 'management'. But, as things seem to get worse, not better, we should reflect that we have had 30 years of courses and development programs dominated by 'leadership' and a parallel diminution in attention to good old-fashioned managing.

He goes on to say that '…leadership came to stand for the pursuit of dreams or visions – and management for the bureaucratic nuts and bolts. Leadership was seen as good. Management, at best, was neutral – maybe necessary'.

Mant then refers us to different facets critical to viewing the essence of leadership – the *transformational* and the *transactional* behavioral facets.

The *transformational* facet – has 'everything to do with destination. The effective leader needs to be smart enough to identify a better destination and to plot intelligent pathways to get there'.

The *transactional* facet – recognizes 'that leaders are ineffective without followers, so the ability to trigger the human following instinct is always necessary, but not sufficient' by itself.

> Good leaders will 'always balance the managerial aspect (running things properly) with the leadership aspect (locating the enterprise in a useful external place)'.

We will examine this model in more detail, but before we do let us have a brief look at why process leadership is important.

Why is process leadership important?

No matter what strategy, process governance and performance, technology, people capability or project execution is in place, if there is no passion for the sustainability of having a process-focused organizational culture or environment then the organization will fail to be successful in the long run.

Davenport (2004) described it as follows: 'Actions are important, but they will not happen without a culture focused on process improvement. Many of us may take such a culture for granted, but rest assured that there are plenty of Neanderthal executives who think that cracking the whip is the answer to their performance problems'.

Case study: Polishing an Ivory Tower

An organization had recruited an ambitious Business Process Manager. His job was to provide the newly established Centre of Business (Process) Innovation with the required tools, techniques and methods to support the business to achieve benefits through process improvement and management. He went to the business directors and explained the services that his group would be providing to them. He and his team started preparing a complete toolkit for business process improvement and management using industry best practices. However, he was unable to get any traction in the business. Recognizing this problem he started putting even more emphasis on completing and upgrading his deliverables. This led to frustration among the business who did not want to use the tools and templates as they did not suit the style of work being completed. This created frustration for the BPM manager who became bitter about the organization and cynical about its process improvement programme.

Message: It is important to understand the culture and maturity of the organization and armed with this knowledge, set targets and instigate appropriate initiatives. Working harder in many cases is not the solution, working smarter is. Listen to your customers (the business) and understand the business needs.

Key trends

We have observed the following trends in relation to process leadership:

- Process-focused culture is progressively being recognized as a corporate asset. The matching of business processes and the culture surrounding them are being taken more into account during the establishment of partnerships and the exploration of mergers and acquisitions. This is a logical result of the many failed acquisition attempts to achieve synergies which were purely focused on the 'cold hard facts' of products, services, markets and geographical location of the businesses, while ignoring the 'softer components' of people change management and a business process improvement culture.

- An increasing number of organizations want to achieve evolutionary continuous business process improvement rather than the more revolutionary business process re-engineering approach of the past. It is our assessment that the reason for this is because executive managers believe that continuous business process improvement is more manageable, less disruptive and the results are more sustainable. This, however, is not always the best approach. Some organizations require an innovative step approach to business process improvements.
- The power of having robust and standard business processes is being recognized as a benefit as organizations become more global due to the opening of international markets, globalization of the society and organizations and the continuous migration of skilled labor throughout the world. The positive element is that organizations can benefit from best practices from around the world. The issue to watch is that as different countries have different cultures and values, the suitability and impact of the current standard business processes need to be seen in this light.
- Innovation can no longer be limited to big breakthroughs by specific research and development departments and must become more systemic within organizations. Innovation in this context includes:

  - the creation of new products and services; new applications for existing products and services and new ways of producing these products and services and
  - the redesign or innovation with regard to business processes. This can include: innovating processes for distribution channels; sales and marketing processes and the traditional 'back-office' processes.

Organizations realize that appreciating a series of small ideas might result in big achievements. Furthermore, a culture fostering process innovation will eventually also lead to breakthrough ideas.

Case study: A tough leader brought to his knees by the 'pen'

A business manager was recently appointed to turn around the prospects of a business unit that was perceived as underperforming by the executive management. He started to immediately make major changes in the business unit and completely ignored the warnings of his direct reports that the amount of change was more than the employees had ever faced and were able to absorb. Despite the warnings from his

(Continued)

Case study: A tough leader brought to his knees by the 'pen' (*Continued*)

management team he forced the changes in the business unit as he wanted to make his mark quickly. His 'take-no-prisoners' blitz came to an abrupt end when the annual employee survey was published, which showed his business unit dropping significantly compared to the rest of the organization. The score for staff engagement and involvement dropped to a historic low. Further analysis showed that his employees did not see the need to change so quickly and drastically. The business manager then went to the other extreme – he let his staff decide the priorities for the coming period. This resulted in many of his changes being totally nullified and a freeze on any other major changes. In effect, he had lost the ability to make any significant improvement in the foreseeable future.

Message: Staff engagement is critical, particularly if major changes are being contemplated. Engage staff and listen to them.

Note: We are not advocating that organizations become a democracy where all decisions are made totally by consensus. It is manager's/leader's job to make decisions; however, it is important to consider the impact on employees when making them. A good leader knows how to rally the employees by explaining the need for change and particular decisions.

- As employees evolve into 'knowledge workers' and are better educated, this creates a situation where the employees have an increasing ability to assist their managers in identifying and achieving improvements in an organization. However, it also provides more of a challenge for the leaders of the organizations as they must upgrade and fine-tune their methods and approaches to motivate, manage and reward their employees.

Key elements of process leadership

Before we describe the key elements as we see them, we should have a common understanding of the key qualities of leaders and who are the leaders within an organization.

We would suggest that there are several levels of leaders within an organization:

1. Chief Executive Officer (CEO), senior executive or business unit manager who are responsible for the delivery of the organizations strategic objectives.
2. Business Managers, Business Process Managers and Business Improvement Managers, people whose responsibilities are to run the operations of the business (business as usual) and to facilitate change in the organization.
3. Operational Team Leaders, programme/project sponsors, programme director and project manager who are responsible for running the business day-to-day and the particular outcomes of a programme/project.
4. People, this can be staff within the organization or staff from partners, vendors and clients.

A leader is anyone who is capable of exerting influence within the organization, and that is everyone, because 'you cannot not influence' other people.

When reading the rest of this chapter, keep in mind that all these people have the potential to be organizational leaders; it is just their sphere of influence that is different.

In the context of business processes, we believe one of the key elements in process leadership is understanding the difference between *transformational* and *transactional* leadership behaviors. Put simply this is the difference between a leader(s) creating a vision for the organization to aspire to and the day-to-day operational management of an organization. The latter sounds mundane and boring when you say 'the day-to-day operational management of an organization' and yet it is one of the hardest things to execute well and will add significant benefit to an organization.

T*ransformational* leadership behavior results in a clearly described picture of where the organization is heading, and also a description of what the future will look like when it gets there – the culture and environment; the behaviors that are acceptable and unacceptable. Leaders need to model the behaviors and develop trust in their capabilities and values. Courage is at the core of transformational behaviors – leaders need the courage to make the right decisions.

T*ransactional* leadership behavior creates and inspires the people to build a *high performance management* environment. They promote a business process aware culture; demonstrate and grow the trust in the leadership by modeling actual behavior; build robust and honest communications channels; build a structure and environment that promote, innovation; and employ and promote, the 'right' people in the 'right' roles.

Visionary process leadership

In visionary organizations there will be a realization that the future requires a different way of thinking, working and managing. If it does not, it will be overwhelmed by the factors outlined in the beginning of this chapter.

The visionary organization will have innovative and end-to-end process thinking embedded throughout the entire organization. All business processes will be geared towards the strategic objectives of the organization and any problems will be reviewed with a systems approach and structural solutions will be sought.

The entire organization will have adjusted to the same process-focused way of working. People working in teams are able to inspire others who have not yet been exposed to an innovative way of working.

There is an emphasis on *management* training and coaching enabling a high performance management environment.

All organizational stakeholders will:

- Have a clear understanding of acceptable and unacceptable behaviors.
- Have trust in the leadership.

Leaders will model the required behaviors and values to all organizational stakeholders. To do this they must be courageous in their decision-making. Therefore it is fundamental to the success of the organization to have the right people in the right roles.

Roadmap to process leadership

Although we have described the visionary process leadership position, the reality is that few organizations have this type of process leadership across the entire organization at all levels. Organizations will vary considerably as to where they are on the journey. There is, however, a growing awareness that 'process leadership' is a critical part of what an organization should be doing. So how does an organization get from where it is today, towards the visionary state?

In the introduction we described *transformational* and *transactional* leadership behaviors. We stated that:

- The *transformational* facet – 'has everything to do with destination. The effective leader needs to be smart enough to identify a better destination and to plot intelligent pathways to get there'.
- The *transactional* facet – recognizes 'the ability to trigger the human following instinct is always necessary, but not sufficient' by itself.

We also stated that good leaders will 'always balance the managerial aspect (running things properly) with the leadership aspect (locating the enterprise in a useful external place)'.

Let us now look at this model in detail (Figure 4.2).

Transformational leadership behaviors

This is what we traditionally think of as leadership. The leader will identify a 'better place to be', create the vision, describe it for us, sell it to us and 'take us there'. In this instance, and because of the topic of this book, we are suggesting that this 'better place' should be a process-focused organization.

However, creating a vision and describing it is a relatively easy thing to do. All it takes is ink, paper and a little bit of thought. The extremely difficult part is getting thousands, or tens of thousands, of people to take the journey and behave the way that the destination requires. It is also challenging for the leader to identify and take into account the organization's capability (maturity) of arriving at this 'better place to be'. The options may be restricted by a low level of process maturity.

This is where true transformational leadership behaviors comes into play. It is where the leader(s) clearly understand the process-focused behavior they must exhibit and then behave that way *all* the time. Leaders need to model the behavior required and clearly make known what behaviors will and will not be tolerated within the organization. All the training in the world will not

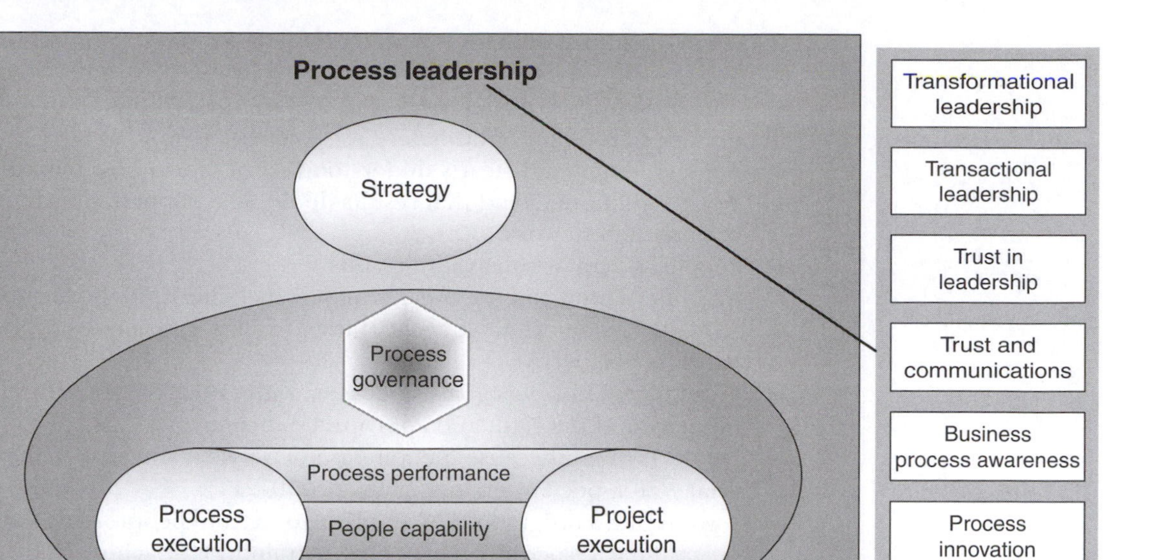

Figure 4.2
Management by Process framework: Process leadership steps.

change people's behavior unless they have role models. This is especially true where the desired behaviors described in the training are incongruent with the behaviors being exhibited by the leaders. It is like parents telling children not to behave in a particular manner, when they are themselves.

> You can change the behaviors of the people within an organization by changing the behaviors of its leaders.

The leader's next most important job is the execution of the plans; the detailed work; making sure staff achieve the desired results and this *cannot* be delegated to others.

Leaders must be intensely and intimately involved with their people and business operations.

As with most activities within an organization, communication is a critical aspect. The core focus of communication to staff and management must be candid (honest); reality-based (and relevant); ask and seek answers to questions and must encourage working together to find realistic solutions.

While many operationally based staff within an organization may find the organizational strategy very complex and high level, and some are, the strategy needs to be brought down to a (hopefully small) number of specific actionable steps that can be implemented and 'made to happen'. Staff must not only understand the strategy and what needs to be implemented, but they also need to believe that it is capable of being achieved.

The leader(s) job is to create an *execution* culture where staff 'get' what they have to do and have an environment where they can 'do' it.

This execution culture will have:

- Embedded performance targets which are directly linked to appropriate and relevant rewards.
- A documented and clearly understood set of 'norms' for behavior.
- People will understand that results do not just happen, they require hard consistent work.
- People will equate ideas with results.
- An understanding of the organizations capabilities and shortcomings.
- A 'habit' of always looking at things from a customer/suppliers/ partners perspective.
- Carefully set business goals or targets, with enough resources to fulfil them, and the right and appropriate incentives.
- Follow through because 'people do not do what you *ex*pect, they do what you *in*spect' (Gerstner, 2002, p. 210–211).
- A continuous programme(s) in place to expand people's capabilities via training (especially management training), coaching and mentoring, identification of talented people and placing them in the right positions. We have seen organizations within the same industry perform 20% more effectively than others largely because the leaders of the high-performing ones motivate and coach their people.
- A culture that does not tolerate consistent bad performances and a process of quickly informing people to enable them to make adjustments, or have the situation dealt with.
- A policy of, and understanding that by, employing talented people in the right jobs will be self-perpetuating because talented people employ other talented people.

Leaders have a huge opportunity to create wealth by 'reducing unproductive complexity and increasing productive interactions' within their organizations (The McKinsey Quarterly, 2007, p. 23). As McKinsey's outline in this article, the 'numbers rapidly become large. If a company with 100,000 employees makes internal organizational-design changes that add $30,000 in profit per employee, for instance, that would mean $3 billion in extra profit'.

Even if we scaled back these numbers to 10,000 employees and $10,000 additional profits each, then the additional profit is still $100 million per annum.

Creating and maintaining this execution culture takes courage from the organization's leader(s) and unfortunately with the pressure on leaders to perform, every quarter, and the politics involved within most organizations, this courage is unusual and yet the benefits are enormous.

We will now discuss the transactional leadership aspect and explore how leaders may be provided with the confidence in their management team to be able to implement this execution culture.

Transactional leadership behaviors

Simplistically, *transactional leadership behaviors* is about running an organization to a high level of operational excellence. It is about putting all the things that are necessary in place to enable the organization to service and satisfy its customers (to a high standard); to have the business processes running efficiently and effectively; to have an outstanding relationship with suppliers and business partners; to continually innovate; and have staff enjoying their jobs so much, that they love coming to work.

For this to occur, leaders need to create an environment within their organization that will perform to a high standard and deliver upon the organizations strategy. We have called this the creation of a *high performance management* model as shown in Figure 4.3.

If an organization wants its managers to perform to a high standard, they need to understand that many (or most) managers will need to change their current method of managing. In fact, the entire organization will need to change and this can be extremely *confronting* to both the organizations leadership and management staff. How can this not be confronting when managers have been executing their roles for decades using a certain set of skills and activities, and they are now being told that it is not working to the expected standard, they must change and they have no clear way of understanding how to change?

No one except the leader and the leadership team can provide the *courage* for the necessary changes to occur. The leader needs to provide the vision of

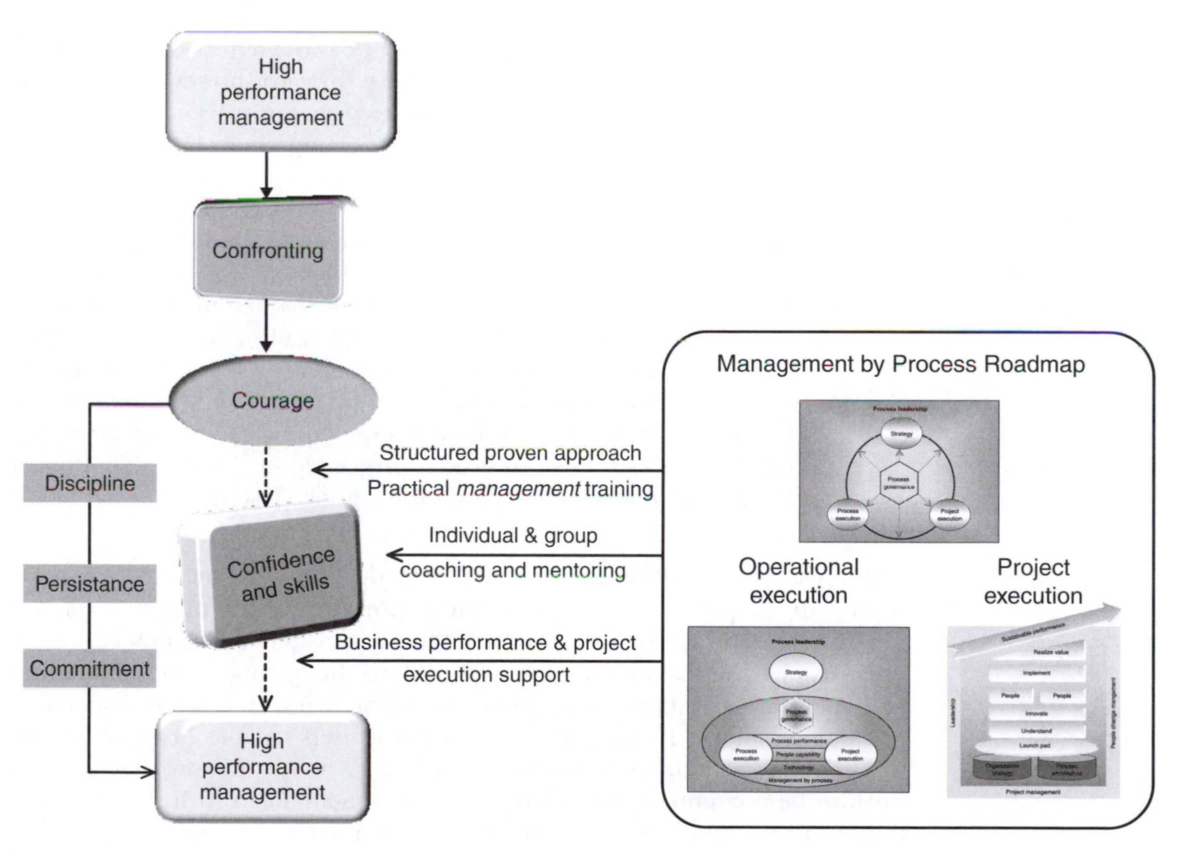

Figure 4.3
High performance management model.

what he or she wishes the organization to look like, behave like and aspire towards (part of the *transformational leadership* aspect discussed earlier).

Not only will this be *confronting* but managers, as they commence the movement to the new high performance management model, may also lack the necessary *confidence* and *skills* in how to achieve what is required of them in a structured, repeatable and sustainable manner. Without both the *confidence* and *skills* high performance management is unlikely to be achieved.

Managers will need to be supported through this transition by:

1. Being provided with the support, encouragement and necessary resources by their leaders.
2. Being provided with a proven structured approach, together with a roadmap and guidelines.
3. Practical *management* training – Most of the training being provided to organizations now and over the last 30 years has been *leadership* based and yet little, if any, has been provided on *management*. This lack of management training is not just within an organization to its staff, but also with more formal training at universities. As Mant suggested, leadership is glamorous (and yet very few of us will ever actually become the leader of an organization) and management is mundane and boring (and yet most of us will be, or are already, managers).
4. Coaching and mentoring – Training provides a necessary and excellent foundation from which to move forward; however, most of us have been on many training courses over our working or personal lives and most of the training has not changed our lives or provided us with a deep understanding of a topic. The investment in training will mostly be wasted, especially in a management setting (how to be a better manager), unless it is followed up and supported by individual and group *coaching* or *mentoring*.
5. In the execution of both day-to-day business operations and business process improvement projects, the organization and man-agers will require a structured approach that minimizes risk, will be successful, repeatable and sustainable. This can be provided, at the day-to-day business operations level, by the seven dimensions and approach outlined in this book and at a project level, by the 7FE Project Framework described in our previous book (Jeston and Nelis, 2008).

This environment will only be created if leadership gets the people aspects 'right'. In Chapter 8 we will discuss the importance of the people within an organization and the support they need to be provided. As we enter times of great change (globalization, global migration, the ageing of the population and therefore significant changes in the labor market), people sustainability is increasing in its criticality to business leaders. At the heart of people sustainability is the perception that people have about the way they need to work to be recognized, rewarded and to ultimately fit in with the organization. Get this right – make the organization attractive to people, have people connected to the purpose (strategic alignment) and have passion espoused by the leaders – then you build people sustainability.

So why then are many organizations still failing to succeed with people sustainability? Many continue to incur high staff-turnover rates, lower productivity rates than they would like, low engagement levels with their people and many other measurable symptoms. The answer is generally quite simple – this failure is largely attributed to the top leadership group who are failing to recognize the importance of matching their own behavior to these values and the expectations set when espousing what the organization should look like.

> Herein rests a critical foundation of trust in leadership. People within the organization want to see actions matching words and then and only then can they have trust in leadership.

Trust in leadership

So how can organizations and its leader(s) provide the required trust to the various staff, management and other stakeholders? The first thing is to identify all the stakeholders to whom trust needs to be won.

There are more stakeholders to trust in leadership than the people within the organization – there are customers, shareholders, suppliers, partners, potential investors, the Board of Directors, lenders of debt finance (such as banks) and they all need to trust the leadership of an organization. The trust they look for is linked to the sustainability of the organization's finances, people and operational capability. These stakeholders look to business leaders to successfully manage this on-going tension within the business. This tension permeates every decision they make, each behavior they choose and every promise they make.

Key stakeholders need to know that the leaders of an organization can be trusted for their ethical behavior and quality of decisions. They need to be able to trust their business conscience and that they will 'do the right thing'.

At the heart of the financial sustainability of an organization is the capability, and willingness, of its leaders to make sound business decisions. Key stakeholders want to develop trust in their leadership to make the 'right' decisions. They want the leadership to show they are insightful, practical, resourceful and lastly decisive. Financial success is paramount – it is measured and reported to key stakeholders.

To develop trust that the leadership decisions will be 'right', these key stakeholders need to see and feel congruence with the values of the organization and the values of the person in the leadership role. People know what to expect because of the espoused and published organizational values, the promises made are actually delivered to delight customers, the financial forecasts made public and the interactions with people in the organization. For there to be trust in leaders, key stakeholders need to see actions matching these promises and matching the personal values of the leader.

So, what are the key issues we need to consider when building trust in our business leaders? In building such trust we need to see:

- Courage to make the 'right' decisions.
- No office politics and self-centred behavior.
- Leaders 'walking the talk' because staff will identify incongruent behavior instantly and deem this to be acceptable and tolerated behavior – 'so we all can do it!'

- Leaders admitting they do not have all the answers, and including their 'team' in the decision-making process. Showing the team that they value their input.

Bad experiences and a driving need to protect ourselves as we move up the organizational hierarchy and to protect the organization have driven our business leaders' to have a low propensity to trust. If this continues then key stakeholders will continue to have low trust in an organization's leaders. It is no different with business staff who must come to trust their leadership if we are to have sustainability in both the organization and the staff. Changing this is vital to the future sustainability of organizations.

Trust and communications

The final aspect of building trust in an organization's leaders has to be in the leader's ability, capacity and courage to communicate their decisions – very openly and very honestly without holding back information. Unfortunately the 'keep them in the dark' syndrome is alive and well throughout many organizations. Decisions are made without consultation and then without sufficient communication and buy-in for the necessary stakeholders. This is a huge reason why key stakeholders (e.g. unions) do not trust organizational leaders – why our initial reaction is to question the ethics, integrity or correctness of a decision when it is finally communicated.

Communicating decisions involves a strong element of vulnerability and this is where many business leaders opt out as it is easier and safer to keep the knowledge to themselves and either:

- 'Leak' the decision over time.
- Share it in a quiet and less overt way.
- Make a brash and commanding statement that has the undertone of 'do not question me or my decision'.
- Let people find out for themselves as they experience the impact of the decision (such as a new policy or system).

Why do some leaders become more untrusting of people as they progress in seniority of leadership roles?

This is a challenging question and perhaps the answer lies in people's fear of being courageous, vulnerable and really exposing what they think and feel. Most leaders prefer to 'keep their cards close to their chest' and empower their ability to negotiate. When you hear the term 'negotiation' being used in conversation then you automatically know there will not be trust! Trust in our organizational leaders is at an all-time low and we are trusting our leaders less and less.

> External and internal stakeholders want the business leadership to be reliable and proven – they want to know the decisions that will be made can be relied on and are trustworthy. They need to know that the leadership will act with integrity and will always be honest – then they know they can have trust in the organizations leaders.

Business process awareness

Now let us get back to discussing leadership from a business process perspective. We were recently speaking to a CxO level executive at a large organization and he stated: 'we are in a mature market and we are a mature organization and do not see much opportunity for improving our business processes'. This is clearly an executive who has not seen the 'shift' happening that was outlined at the beginning of this chapter. It is interesting how many organ-izations predominantly only see opportunities for profit growth from an external perspective – either increasing sales by taking market share from competitors or acquiring another organization.

Few organizations look inside. As Hammer (2004) states:

> it should not be surprising that executives without experience in operations do not look there for competitive advantage. The information they usually get does little to focus their attention on the mechanics of operations. How many executives receive data about order fulfillment cycle time, or the accuracy of customer service reports, or the cost of each procurement transaction, or the percentage of parts that are reused in new products? Indeed, in how many organizations is such information available at all? Financial data dominates the discourse in the modern organization, although operational performance is the driver of financial results.

Leaders and managers have rarely been exposed during their formal training to the impact of business processes on an organization. Many managers are promoted because they are good at solving problems ('band-aid management') rather than their ability to get to the bottom (root cause) of the issue and fixing it permanently.

Citibank Germany (refer to the case study) was not initially successful at creating an awareness of the importance of business processes within the organization. However, they kept trying until they found a way that worked for them. Unless this can be achieved, then the chances of moving forward to a high performance management model will be significantly diminished. Business performance, at the operational level, predominantly relies on the performance of business processes.

The challenge for us is not so much about having the people within an organization understand the importance of business processes. The challenge is for the senior executives or leaders to 'get it' *and* have the *courage* to take action. Show your leaders the case studies in this book. Find other case studies and show them. Without the support of leaders, we will all be delegated to small business process improvement projects that are either 'under the radar' of managers or perpetually in 'pilot' phase. While either of these approaches can provide a start, even when they are successful, many times, leaders often still lack the courage to take them further.

Process innovation

If an organization wants a high performance management environment, then it will continually need to question the way it runs its business. Process-focused leaders will create an environment where innovation is part of what the organization does. We suggest that there are three types of innovation (Dundon, 2002):

- *Efficiency innovation* – improving the current business processes and systems for the existing products and services. Most of the continuous business process improvement projects focus on this area. This particularly relates to faster turn-around times, lower costs, better quality and more control.
- *Evolutionary innovation* – identifies new and better ways to bring value to an organization, by finding new ways to improve the current products and services and reaching other markets. Business processes can be a strong enabler for this, for example, Dell's approach to changing its business processes to enable customers more flexibility and control.
- *Revolutionary innovation* – involves the introduction of radical new ideas that affects not just the organization but the entire marketplace. This often leads to new benchmarks in the industry. Business processes, combined with technology, can bring those changes, for example, the on-line booking of tickets has reshaped the entire travel industry.

Organizations need to understand that innovation of its operational areas, especially if *revolutionary*, is by nature disruptive. In selecting the approach and the business processes to innovate, leaders should concentrate on those with the greatest impact on an organization's strategic objectives.

Case study: Significant innovation and leadership impact

We provided some consulting to a large energy organization because it needed to improve customer service. Property developers and similar organizations would approach the organization to provide power to a new housing estate or shopping complex. The current high level business process was extremely reactive for the energy organization which resulted in many disputes with its customers (the developers). A totally new approach was developed to redesign the high level business process to enable the organization to take a more proactive approach and provide significantly better customer service.

When the new business process was presented to the general manager for approval, he clearly stated that his electrical engineers were not skilled enough to provide this proactive approach and rejected the idea for two years.

This reaction reflected the general manager's approach to his staff. Customers knew they simply had to complain directly to the general manager and he would consistently override the decisions of his staff in favor of the customer, even though this placed the electricity grid in long-term jeopardy.

Message: Had the organizations business leader (the general manager) worked with his staff and trusted them (even if they needed coaching and training) the organization could have moved forward to be a higher performing organization.

Many organizations are still trapped in Taylorism, where the 'brains' of the organization are separated from the 'hands'. That is, a limited number of managers and subject-matter experts determine what the 'workers' should be doing and how they are doing it.

These types of organizations face the following problems:

- The managers have limited insight of the actual processes.
- The staff will not be engaged and therefore may not be willing to implement new ideas.
- Staff will not provide ideas and suggestions for innovation and business process improvement, which results in a lower level of performance improvement.

A critical change in attitude is required, namely the engagement by managers of the staff and a willingness to take risks, which opens the possibility of failure. However, as Einstein is purported to have said: 'anyone who has never made a mistake has never made anything new'. Over-controlling managers need to be assisted and coached so that they themselves can guide and coach their staff. Innovation will only come when staff feel safe that their suggestions will be taken seriously.

However, innovation within an organization just does not suddenly arrive by itself. Nor does it arrive by the sudden generation and use of slogans or the recruitment of 'clever' people. Innovation needs to be nurtured and facilitated, and sufficient time needs to be allocated to allow it to emerge.

Google is an innovative organization and requires all of its employees to spend 20% of their time working on any project or idea of their choosing (Google, 2007). Google services such as Gmail and Google News started as one of these 20% ideas. The ideas that people generate during this '20% time' will often have nothing to do with Google's current core business. This provides an environment in which fundamentally different services and products may be generated.

Not all organizations need to allocate 20% of their staff time to innovation. It all depends on how the organization wants to position itself. The more innovative it wants to be, the more time it needs to allocate and provide the necessary facilitation and support to use this time effectively.

Other ways of being innovative include:

- Joint study programmes – a popular approach is for a group of subject-matter experts, leaders or potential leaders to complete a management course together, either in-house or at a university. The assignments should be completed as a group and relate to the specific circumstances within the organization. The outcomes of these studies could be presented to senior management who may choose the most promising proposals for further pursuit. This approach will promote teamwork among the participants.
- Award programmes, idea campaigns and contests – triggers employees to provide ideas and concepts. The challenge is to have people think outside-the-box and come up with ground-breaking innovations and ideas.
- Sabbaticals – provide employees with a break from their day-to-day work to provide further education and training which may provide

staff with an opportunity for some thinking and inspiration time. However, although a sabbatical provides employees with a well-deserved break from their normal work, it does not necessarily provide innovative ideas.

- Job rotation and internships – provides employees with the chance to experience different views and aspects of an organization. This can be especially powerful when people are brought into a business unit which has a track record for innovation and/or process improvements.

Case study: Innovation as service offering

We participated in an organization's innovation workshops. This involved a facilitator who had been especially trained. Additional facilities were also required to allow the right atmosphere for innovation. Managers and staff were invited to discuss specific problems, challenges or opportunities. Workshop duration varied from two to eight hours. Participants were taken outside their comfort zone and encouraged to think outside-the-box, through music, games, role-plays, etc. A wide range of diverse ideas and suggestions was identified. The ideas were then consolidated and aligned with their purpose. Finally concrete actions were specified to ensure that the innovation was executed correctly. The success of this incubator scheme and innovation room became so successful that customers were also invited to participate to obtain even better ideas and it also indicated the commitment of the organization to providing an outstanding level of service and satisfaction to its customers. Eventually, the customers were so impressed they requested the use of the innovation incubator facility for their own businesses.

Message: Innovation can provide a competitive advantage if accessed and implemented in a systemic way.

While brainstorming sessions can be a useful way of deriving new ideas most organizations do not have mechanisms or processes in place to follow-up and cash-in on the ideas generated. This eventually leads to disappointment and frustration for the people who provided the idea in the first place and hence a lower level of enthusiasm for future brainstorming events. To overcome this problem we suggest the following phases:

1. *Idea generation* – the purpose of this process is to obtain as many innovative ideas as possible.
2. *Idea processing and fitting* – the purpose of this process is to screen the ideas and assess their feasibility and compatibility with the strategy and objectives of the organization.

   The following approaches can be used for this purpose (Souder, 1988):

   - Ideas inventories: having a running inventory of ideas – this can provide a wealth of information and inspiration. The organization must determine how much information is required when an idea is submitted and accepted.
   - Idea clearing house and brokers – a matching facility is provided to match and forward ideas which meet the specified needs of various parts of the organization. They might also try to find appropriate solutions for emerging problems and challenges.

- Idea fitting teams – this team tries to adjust existing ideas to the specific needs of the organization. It is advisable that the submitter of the idea is present when his or her idea gets 'fitted'. This provides them with an opportunity to contribute to the 'fitting' process. Furthermore, it also provides the submitter with a better insight into how ideas fit within the organization. This can be beneficial for future ideas.
- Idea screening and review teams – this team assesses the relevance of an idea early in its lifecycle. This ensures that the provider of the idea has a fast response and avoids the situation where unnecessary time and resources are wasted on inappropriate ideas.

A key role during this process is to have people with entrepreneurial skills – people who can recognize the potential of ideas – to work closely with the people who submitted them.

3. *Acceptance and Commitment* – the purpose of this process is to formally accept an idea and commit the required resources to it so that it can be implemented. In many cases the submitter of the idea might not be in charge of rolling it out within the organization. It is important that clear ownership of the idea and roll-out is assigned to ensure that an idea achieves its full potential.
4. *Idea Portfolio Management* – this is a systems-based portfolio and has the following benefits:
   - The formal ability to track the progress of ideas.
   - The formal ability to track the benefits of the ideas.
   - Provides a report on the overall portfolio of ideas and their status and progress.

Thinking outside-the-box

We have mentioned several times thinking '*outside the box*'. This popular saying separates *inside* the box thinkers (the vast majority) from *outside* of the box thinkers.

In-the-box thinkers are 'stuck' within the current system and way of thinking. They find it difficult to recognize the quality of an idea as they prefer the prevailing status quo. To them an idea is an idea and a solution is a solution. In general, they would assess (or assassinate) ideas on the basis of their impact (disruption) to the current system.

Another characteristic of *in-the-box* thinkers is that they believe there is only one solution for each problem. Thus, they stop with the first idea that addresses the problem and find exploring other (possibly better) solutions a waste of time.

Outside-the-box thinking requires a different thought process that includes:

- Looking from a new perspective at common tasks and day-to-day work.
- Openness to try different things and to do things differently.
- Readiness to fail in certain endeavors, as not everything can be successful, but ensuring they learn from the mistakes.

- Focusing on the value of finding new ideas and acting on them.
- Listening to others and exploring new ways together.
- Supporting and respecting others when they come up with new ideas.

Once a good idea has been generated and examined, many people think that it will just sell itself and that everyone will understand and appreciate the value that the idea or innovation will bring to the organization. This lay-back approach can, and usually will, lead to disappointing results when the idea is not automatically adopted.

Ideas need to be sold and justified just like any other activity within the organization. It may require a business case or a feasibility study. Just remember, if you are passionate about the idea, do not give up.

Remember, that 'operational innovation may appear unglamorous or unfamiliar to many executives, but it is the only lasting basis for superior performance. In an economy that has overdosed on hype and in which customers rule as they never have before, operational innovation offers a meaningful and sustainable way to get ahead – and stay ahead – of the pack' (Hammer, 2004).

Promoting the 'right' leaders/managers

The last area we will examine as part of process leadership is making sure we promote the right people into leadership positions. It is vital to have the 'right' people being promoted into the 'right' jobs. Leadership, being at the forefront of a business, a project or team, has very little to do with a persons technical mastery, and everything to do with who a person knows, how they behave, how they engage and motivate those they work with and who needs to follow them.

People develop guru status and acquire a strong professional reputation as they develop their technical mastery. It is then all about others knowing that they develop products, services and solutions that deliver value. Think of an engineer, an accountant, an architect, a lawyer – they spend their study/qualifying time developing technical mastery. They use their skills to advance their career. Then it is their 'guru' status that gains them their right to become a business leader. They were trusted because they could be relied upon to produce an outcome that was needed – that is, to build a reliable and aesthetically pleasing building, or to produce a set of reliable and correct financial accounts, for example.

One of the biggest mistakes organizations are making is their continual practice of promoting technically competent people into leadership/management roles. This can actually create distrust all-round because:

- The person themselves begins to lose trust in their own capability – they used to be respected and rewarded for their work, now they cannot seem to get a group of people to follow them. Their self-trust is broken.
- Those responsible for the appointment lose trust in the person – they used to be good at what they do, but now they cannot make a business decision to save themselves? The trust the shareholders or directors had in them is broken.

- Those expected to follow lose trust in them – they might have been good at what they did, but now they have not got a clue about how to build relationships with people. The trust their followers had in them is broken.

Key stakeholders learn to 'respect' the business leader because they have strong technical mastery; they can trust the solution and the advice – but they do not have trust with the business leader as a person.

If we were to leave you with two critical thoughts from this chapter on process leadership it would be these:

From a process and operational perspective a leadership's 'real' job is about execution. This requires the *detailed* creation and on-going maintenance of an execution culture within the organization. While we acknowledge that this is difficult for leaders when they are under constant pressure to perform, every quarter, and with the politics involved within most organizations, and yet it is *not an option.* The long-term sustainability of the organization demands it.

For this to occur it will be extremely confronting for many of the senior and middle managers with the organization. Without the ceaseless drive of the leader(s), the execution culture will not occur – this ceaseless drive takes *courage* which, unfortunately, far few leaders have. You can always spot the courageous leaders – they can be recognized by the outstanding success of their organizations. Look at many of the case studies outlined earlier.

Chapter 5

Process governance

Introduction

Process governance is arguably the most important dimension for the continued sustainability and long-term success in creating a process-focused high performance management organization. Figure 5.1 shows how it links with the other dimensions. Without a level of leadership commitment to the establishment of a governance structure based around business processes, it will be extremely difficult to achieve anything except isolated business process improvement (BPI) projects and activities.

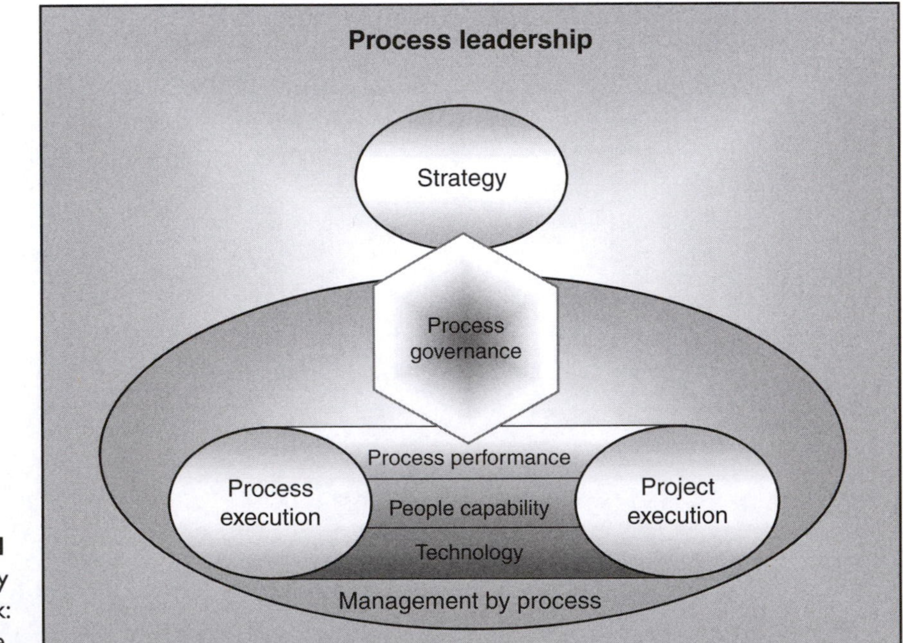

Figure 5.1 Management by Process framework: Process governance.

This is not to suggest that an organization may not be successful in process improvement projects and activities, it is just that these projects will probably be more isolated, not necessarily co-ordinated or strategically focused. These projects may solve individual process problems (bottlenecks, quality issues, throughput constraints and so forth) but are unlikely to lead to overcoming the challenges associated with the resolution of the traditional functionally based silo organization.

Why is process governance important?

Traditionally organizations define governance similar to this statement: 'It is about controlling your business and all its moving parts so your goals can be met and value created for shareholders without violating any laws, covenants with lenders and partners, and without exposing your company's assets to unnecessary risk' (Align Journal, 2007, p. 9). While we agree with this statement, we believe governance should be defined in a much broader way, in the context within which we are discussing it.

If someone is not setting the 'rules' of behavior for the organization, then do not be surprised when people do not play by the 'rules' and make up their own 'rules'. Process governance is about establishing these 'rules' so that the organization may move forward as one, with a level of management, control and synchronicity that it has not achieved in the past.

It is about having executive leadership, management, staff, business processes and supporting technology (IT) all moving in the same direction supporting and building from each other. It is about leverage. It is about team work, where TEAM stands for Together Everyone Achieves More. It is about 1+1=11, not 2. It includes the measurement and management of the organization on a day-to-day basis – this process performance is so important we have dedicated a separate chapter to it (Chapter 6).

While this seems obvious and common sense, it simply does not happen in many organizations. Management and staff are either focused on the best outcomes for their business unit or division (or themselves) and this is not always the best business outcome for the organization or for the customer.

While process governance, and indeed organizational governance, is about risk management and regulatory compliance, it is also about increasing an organization's efficiency, measurement, accountability for decisions made and the alignment between the major components (e.g., business strategy, business processes and IT).

Governance is not just about the formal legal requirements; it is also about the perception in the market place, and is especially focused on the following four groups:

- *Shareholders* – the value of the share-price will increase depending on the level of control the executives have over their business processes and their ability to eliminate adverse business process outcomes. Adverse business process results will have an ever-increasing impact on the profitability of the organization and eventually its sustainability.
- *Customers* – the overall image and perception that customers have about an organization and the customer's willingness to buy products

and services from that organization. This also relates to the perception of the on-going service and support provided after purchasing the initial products and services.

- *Partners* – the degree to which partners are willing to be associated with the organization and the degree of trust they put in the organizations ability to meet its commitment in a timely and orderly fashion. This is particularly important as more and more organizations become dependent on its ability to select and manage its partners and contractors. Governance comprises the structures, policies and procedures by which the organization manages itself. Its purpose is to establish a governance framework to guide investment decisions, provide project and process visibility, assess and manage (or mitigate) risks, all while adding value to the organization's strategic objectives. Without effective governance, an organization is simply operating in a sub-optimal environment.
- *Employees* – the respect and acknowledgment that employees are valuable assets. With the changes in the aging population and globalization, there are many more opportunities for employees, and they need to be respected to retain and grow them.

Effective governance is therefore a critically important dimension for successful organizations.

The Butler Group (IT Governance: Managing Portfolio, Projects, Processes, and People, April 2007) define governance as 'the creation of a management framework by which an organization maximizes the value that it derives' from the various organizational components (IT is only one of them) in supporting the delivery of its strategic objectives. While this report is predominantly referring to IT Governance, the ideas equally apply to process governance. They suggest that a governance framework should include the following characteristics:

- 'defines rights, roles and accountability for decision-making'.
- 'provides visibility' for informed investment decision-making 'and a mechanism for evaluating and prioritizing these requests'.
- 'supports the measurement of value over the' investment life cycle.
- 'defines and enforces standardization' of 'processes'.
- 'helps the organization to assess and manage risk, and to comply with regulatory requirements'.
- 'provides for the efficient management and safe use of resources and assets'.
- establishes various guidelines and frameworks, such as a process architecture, benefits management framework, BPI project management approach and project implementation framework and the project management methodology. These will ensure the consistent, repeatable and sustainable delivery of successful projects.
- performance measurement, recognition and reward structures.

However, it is important to note, the establishment of an effective governance framework is not an end in itself, nor are there templates or out-of-the-box solutions that suit every organization. An effective governance framework is a multi-faceted and complex implementation. It needs careful planning, matching to an organizations culture, continual review, measurement and maintenance. Unless the organization has support

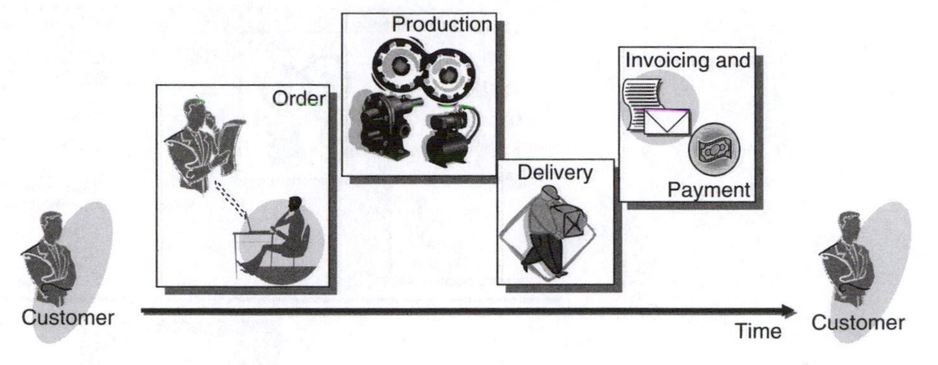

Figure 5.2 End-to-end business process.

from executive leadership, management, appropriate staff and indeed, the organization as a whole, then it will be a significant challenge to make it work effectively.

In the remainder of this chapter, and throughout the book, when we refer to governance, we will be referring to *process governance*, unless otherwise specifically stated. In the following section we will mention the aspects of governance as they apply to business processes and while we describe an ideal visionary way to establish process governance, the reader does need to take note of the comments in the preceding paragraph – there is no 'perfect' solution that suits *every* organization.

So in a simple summary, process governance provides a means for the alignment of business strategy and the high performance management of the organization via its business processes.

Before we begin, it is important to ensure we have a common understanding. First, when we refer to a business process we have used the following definition from Mark Smith (2006):

> A process is a construct for organizing work so it:
> can be ***performed*** effectively and efficiently
> offers the potential for a ***competitive advantage***
> can be ***managed*** effectively.

The only comment we would explicitly add to this definition is that *all* processes must be viewed from a customer and an end-to-end perspective. Figure 5.2 depicts the fact that every end-to-end process must start and end with a customer. Sometimes the customer may be an internal customer, but a customer nonetheless.

Figure 5.3 shows the same end-to-end process from the organization's perspective and the fact that some parts of a business process may be outsourced and *outside* the organization.

Key trends

The key trends we have observed include:

- Many organizations have learned to balance the need for governance, including external regulations and other requirements, with

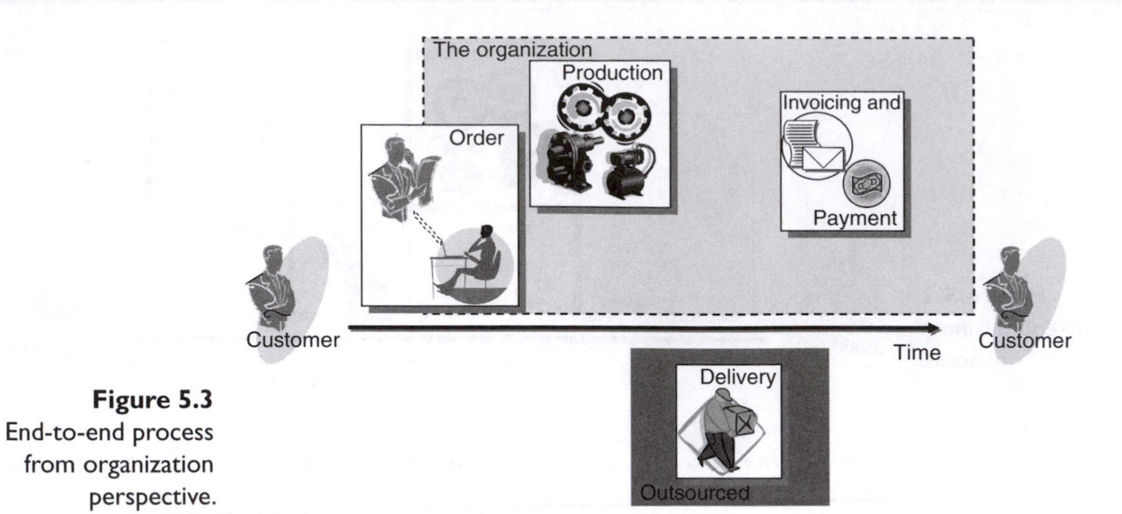

Figure 5.3 End-to-end process from organization perspective.

the need to remain agile. In the past, many organizations have either over or under governed as they have misinterpreted the governance requirements.

- Governance is moving to a more holistic approach rather than separate governance mechanisms for strategy, projects, processes and overall architecture. The Management by Process framework outlined in Chapter 3 provides a useful framework for this.
- Governance is becoming more and more embedded in business and management as usual activities, which can be compared to the position of quality. For many years quality activities were isolated from the actual work; now it is embedded in the business processes with only overall quality management separated from the processes themselves.
- With increasing clarity on how governance is affecting process execution and process management, we expect that standard BPM Systems will support governance functions in the near future.
- Governance is becoming more and more a competitive aspect. It is being used to assure investors, re-assure communities, appeal to customers and make the employees proud. Organizations are starting to make visible to external stakeholders more and more of their governance effort.
- Employees are becoming more aware of governance and all its aspects. They are becoming more able to recognize lack of governance and are more inclined to report non-compliance.

Key elements of process governance

The key elements of the process governance dimension are shown in Figure 5.4 and are as follows: specific roles and responsibilities, roles selection and how to go about it, process management control and business cases.

Figure 5.5 shows a sample process governance framework structure to support business processes in an organization. While aspects of this figure will be part of the ultimate visionary state, other parts must be viewed as part of the

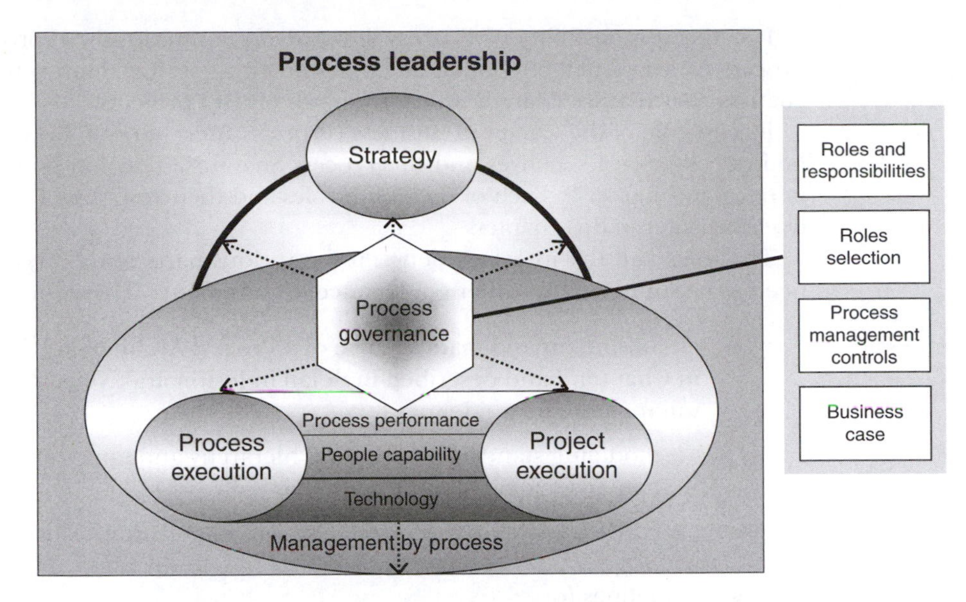

Figure 5.4
Management by Process framework: Process governance steps.

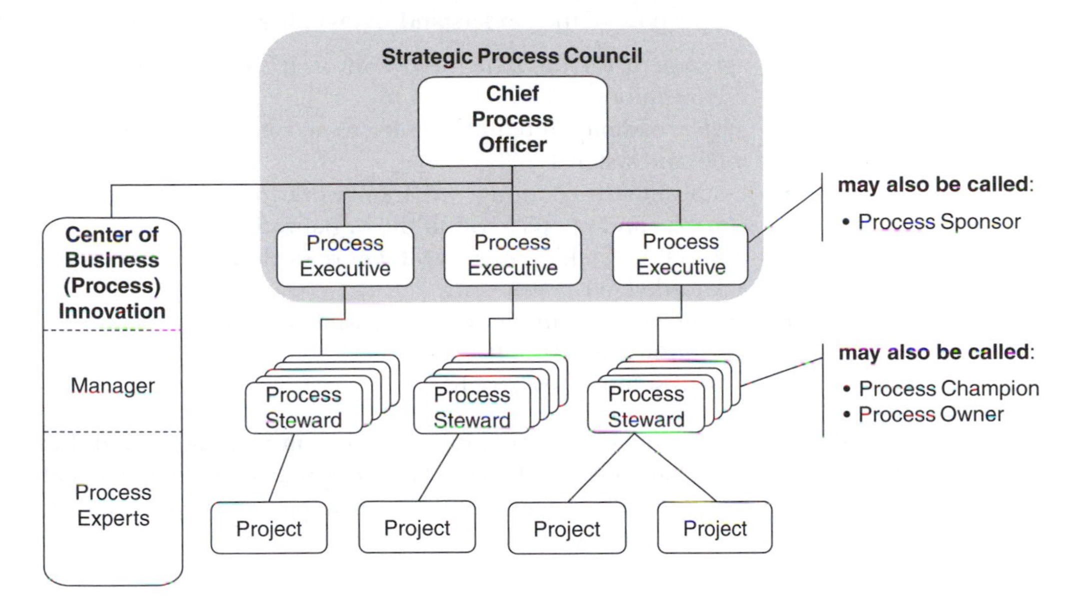

Figure 5.5
Sample process governance structure.

journey and will not be necessary in the visionary state. The Strategic Process Council (SPC) is the primary governing body for an organizations business processes, both from an investment and management perspective. The Chief Process Officer (CPO) is the senior executive responsible for business processes across the organization. The CPO is responsible for the management of business processes through the Process Executive(s), who have Process Stewards (Owners) responsible for individual end-to-end processes.

It is the responsibility of Process Stewards to continuously improve their process(es) via small incremental improvements, or for improvements or changes that require a larger effort, to spawn off BPI projects.

It is the role of the Center of Business (Process) Innovation (CBI) to support both the Process Executives and the Process Stewards. This will be described in detail in Chapter 9. Each of the above roles and their responsibilities will be described later in this chapter.

The roles and the interface between the various parts of the organization need to be supported by other governance requirements. These include:

- establishment and maintenance of a Process Architecture (discussed in Chapter 7 and described in detail in Jeston and Nelis, 2008). This will include the guidelines for:
  - modeling standards, including hierarchy and approach
  - benefits management framework
- executive leadership rules for the 'speed' of, and criteria for, decisions-making
- guidelines for:
  - process metric approaches and criticality
  - project initiation
  - purpose of the Understand (As Is) phase (Jeston and Nelis, 2008)
- agreement for the structured approach to BPI projects within the organization
- an IT roadmap and architecture to accommodate the various BPM tools and systems
- establishment of an agreed framework for the creation of targets (process and people) within the organization
- how target achievement will be rewarded and due recognition acknowledged
- an engagement model for the business, IT and process executives, including the CBI staff
- issue resolution approach.

All this then needs to be communicated in a simple, targeted and effective manner to all stakeholders. It must be noted that some of the stakeholders may be external to the organization, for example, customers, partners, suppliers, regulators and investors.

Visionary process governance

The visionary process governance management framework will create both a structure and set of standards and guidelines to continually move the organization forward, which will lead to the creation of an organizational competitive advantage. The reason for describing a visionary state is to provide a goal for the organization to aim towards.

While the left side of Figure 5.6 shows a reasonably mature process-focused organization, the right side shows the visionary state, which is an organizational

Figure 5.6 Journey and visionary state.

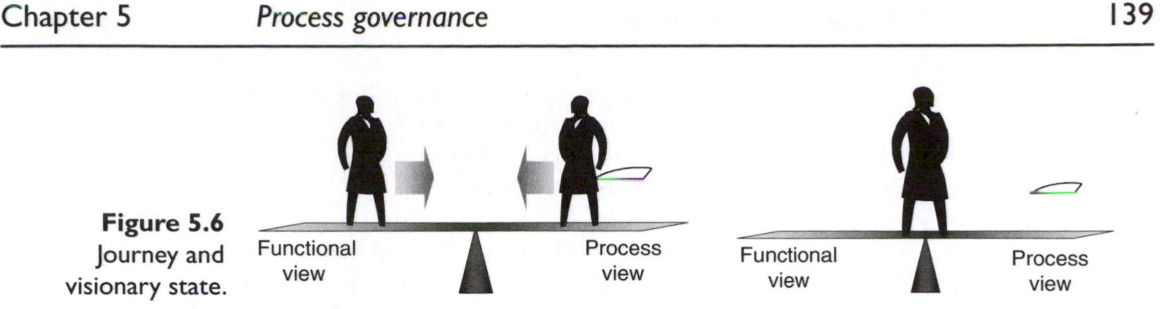

state where process-thinking and focus is 'just what we do around here'. It is an organizational state where executives, management, team leaders and staff all have process clearly in the forefront of their thinking. If there is processing errors, process performance issues, strategic objectives to be met or new technology to be introduced, the organization will clearly understand the impact upon its business processes and know how the business processes will need to be managed.

For this to be a truly visionary state, executive leadership must place their 'stamp' on the process governance management framework. This means not only on the importance of process governance to the organization but also for the various roles, responsibilities and speed of decision-making. If the governance structure and roles and responsibilities are excellent in their construct and still do not deliver decisions with appropriate speed and clarity then significant momentum will be lost.

We will now describe the roles and responsibilities of the SPC and provide a brief outline of the process management controls that are necessary.

Roles and responsibilities

In terms of the sample process governance structure depicted in Figure 5.5 the only aspect that is applicable to the visionary state is the SPC, although even this is arguable. If this council is to remain a separate entity then it will fulfil the roles and responsibilities that are outlined below. If it is not to be a separate entity then the responsibilities will be absorbed and completed by the senior executive council/management meeting of the organization – some organizations call these groups a 'management board'. For the purpose of clarity we have assumed it as a separate governing body within an organization and have described its roles and responsibilities below, although it could clearly be incorporated into, say, a management board.

Strategic Process Council – roles and responsibilities

At arguably the highest level of process governance is the SPC. Note that in our previous book (Jeston and Nelis, 2008) we referred to this Council as the Business Process Architecture Committee. The Council incorporates the Architecture Committee and many other responsibilities.

Composition

This group is typically composed of:

- A CPO who is often the chairperson (note, this role may not be part of a visionary state and only be a transitionary role)

- Process Executives (senior business line executives)
- Chief Information Officer (CIO)
- Chief Financial Officer (CFO)
- Chief Operations Officer (COO)
- Perhaps even the CEO, depending on the size of the organization and the interest of the CEO. Process aware CEOs will add significant influence and direction to the organization through this Council. Do not underestimate the power of this upon other members of management and staff; it will be significant.

Responsibilities

These are wide and varied, but in essence the group is responsible for the key business processes of the entire organization. It should ensure that a central process repository of business process models is established and maintained and that there is a clear link between the organization's strategic objectives and the business process goals. It should understand the processes sufficiently to fully appreciate the extent to which each key process supports one or more of the defined organizational strategic objectives.

> When the strategic objectives of the organization change, this committee analyzes the impact of the change and then directs and changes the business processes to fulfill these new strategic objectives. If this impacts the process architecture, then the changes should feed back for its ongoing maintenance and relevance. Remember:
>
> Alignment of organization strategic objectives and the supporting processes doesn't just happen, it must be planned.
>
> (Jeston and Nelis, 2008)

While the SPC is interested in the business processes, it is not the lower level individual processes or sub-processes, but the higher level core business processes that they should completely understand and know the impact of each upon their business. In detail, the Council is responsible for:

- determining the required level and then the achievement of this required level of process maturity in the organization, relating the current situation, the organizational strategic intent and the benefits to be gained from a process-focus.
- maintaining knowledge of the high level business processes in sufficient detail to ensure that:
  - the organizational strategy can be implemented in an appropriate way.
  - investment decisions are made to maximize both investment spend and organizational capability.
- approval of the Business Process Engagement Model (refer to Chapter 9 for more details).

- initiation, prioritization and governance of organizational key BPI projects, always ensuring that they are aligned with the organizational strategy.
- successful completion of the portfolio of BPI projects.
- ensuring that the central business process repository is maintained and up-to-date.
- any conflicting strategy, business process or investment issues are resolved. This should be capable of being addressed at a lower level (e.g., Process Executive or Process Steward, while on the journey to the visionary state; and by management when the visionary state is achieved), but if this is not the case, then the Council must step in and determine a resolution.
- maintaining business process metrics, for example, the cost and time of each process or group of processes.
- the approval of various business process-based guidelines, such as:
  - the structured approach (framework) to be used by BPI projects (refer to Jeston and Nelis, 2008).
  - how projects are initiated and the impact upon project approaches. For example, is a project strategically driven, business or process driven.
  - project management methodologies (refer to Chapter 9, Project Execution), how they are linked to BPI project approaches and ensuring their use is 'mandatory' within the organization. This again sounds obvious, but we have experienced many organizations which have project management methodologies and they are optionally applied within the organization. It is the project manager's decision to use it or not!
  - approval of the high level approach to the continual 'training' of the organization personnel to be process-focused. The execution of the training is clearly a line management responsibility on a day-to-day basis.
  - process architecture (refer to Chapter 7 for more details) which will include:

    business process modeling guidelines and standards.

    project metrics approach, analysis and how critical this is to the organization.

    benefits management framework.

    project initiation – how projects are initiated will impact how BPI projects are executed.
- standards and guidelines must also be established, approved and continually maintained for the approach to rewards and recognition throughout the organization. We will address this in more detail later in this chapter.

As the Council establishes each of these components, they must be communicated to the organization, and all stakeholders, in a clear and concise manner. There should be no room for ambiguity as misinterpretations can cost the organization a great deal of time and/or money. Those affected directly by any aspect must be dealt with in a clear and caring manner.

Process management controls

Process management controls are about ensuring that the process governance structures, process guidelines and performance measurement targets, and reporting measures are continually up-to-date and appropriate for the organization. These will need to be reviewed on a regular cycle to maintain this currency.

For example, the agreed process-focused meetings established via the process governance management framework should be scheduled on a regular basis. Process management controls is about periodically reviewing these meetings to determine their frequency and appropriateness. The organization must also ensure there are mechanisms in place to ensure that compliance and risk management are in place and appropriate.

Review cycles

The SPC must establish a regular review cycle to continually enhance and grow the various components of the process governance framework within the organization. With experience and growing process maturity, the visionary organization will have learned valuable lessons that must be captured and fed back into the governance framework.

Each component should be periodically examined individually to ensure they are still consistent and synergistic with one another.

The various components that must be reviewed include:

- investment decision criteria and priority setting for BPI projects and activities. We will cover the purpose and use of the business case separately later in this chapter.
- effectiveness of the organization's alignment of BPI projects and activities with the organization's strategy and objectives.
- process performance targets structure and reporting mechanisms, for example, is the Balance Scorecard approach effective, should it be replaced, modified or enhanced?
- if specific process management roles still exist – are they still appropriate or should they change? Note: this is not about the performance of individuals, simply the role.
- the BPI project implementation and structured approach (framework).
- project management methodology.
- process architecture.
- benefits management framework.
- process models – frequency of updating them to reflect the current state.

The frequency of this review will vary from organization to organization; however, we would suggest that each of these should be reviewed at least annually and perhaps more often if there is a particular need. Certainly as significant lessons are learned from management activities within the business and projects, individual components should be enhanced.

All changes must be approved by the SPC. The reason for this is to ensure consistency of all aspects across the organization.

Roadmap to process governance

Introduction

Up to this point we have described a visionary process governance framework for an organization to aspire towards. The reality is that few organizations are at this visionary state when it comes to being process-focused. Organizations will vary considerably as to where they are on the journey. Most organizations do, however, understand that 'governance' is a critical part of what they should be doing (particularly at a board level) and regulations, such as Sarbanes–Oxley, and corporate events such as Enron has further highlighted this from a financial process perspective.

In this part of the chapter, we will examine each of the areas addressed in the visionary section and how an organization may commence the journey towards it. We will separately examine: the various roles and responsibilities; how the people may be selected for the roles; process management controls and the role of the business case in governance. While we examine these separately, it is almost an impossible task because of the significant overlap between each component and their impact on one another, and the complexity of each organizations unique starting point.

Roles and responsibilities

> Imagine that an organization has established the governance structure, placed the people in their roles, and trained them in their new responsibilities. Now comes the hard part, how do we ensure that it all works? Unfortunately, this is not an usual situation within an organization.

Most organizations that start the journey to being more process-focused are initially challenged in establishing the roles and commensurate responsibilities that will suit them at that particular time. The usual starting point is to appoint a process owner. What does 'process owner' mean? First, the role certainly should cover the entire end-to-end aspect of a process. Does this mean they 'own' the process and the users or executors (stakeholders) of the process report to them from a functional perspective; or that the process owner makes all the decisions associated with the process? We suggest that the term 'process steward' is a better description as it implies the true role – that of custodian of an end-to-end business process – and the need to work collaboratively with other process stakeholders to achieve business outcomes. Process steward is the term adopted throughout this book.

In the initial stages of becoming process-focused, the role of process steward will usually be a different person to the functional managers who execute the process from a business operational perspective. This is usually the best approach initially, and it may evolve over time so that one person becomes both responsible for the process and functional aspects of the business.

What roles will be required to start the organization thinking with a process perspective? We will suggest some of these roles and their responsibilities.

Process Steward

The first role created will usually be that of a Process Steward and is perhaps the key process person in the process-focused organization. This is the role that makes process improvement a continuous activity within the organization. It is the role that is responsible for business process performance measurement and the on-going management of the business processes. The appointment of the 'right' people to this role is critical and will have a significant impact on the organization's management of its business processes and therefore business success. We will now describe the process stewards' responsibilities. The skills required for this role have been included in Appendix A. We will discuss the various ways a Process Steward may be selected a little later.

Responsibilities

The responsibilities of a Process Steward are wide and varied and ideally should include the following:

- All process documentation is managed and current, relevant, up-to-date, easy to use, in a central process repository, published, modeling meets all process architecture and compliance standards.
- All business processes are modeled from the customers' perspective. This will include the identification of customer 'moments of truth'. These are the few activities within a business process where, if handled superbly, will significantly emotionally impact the customer and create or deepen customer loyalty and delight.
- All process interfaces and boundaries are managed so that:
  - the customer gets the required outcome from the end-to-end process.
  - details of the end-to-end view of the process are accurate.
  - they ensure a smooth transition between and across process boundaries. For example, end-to-end process problems can arise at the interfaces between other individual processes. The process steward must ensure that the boundaries between various processes are well understood and documented.
  - Service Level Agreements (SLAs) are established and being achieved. An organization must be very careful with the use of internal SLAs. Often when an end-to-end business process crosses departmental boundaries and one business unit or department is not happy with the performance of another business unit, they resort to the establishment of SLAs to raise performance levels. This is not always successful because the inability to meet the SLAs can be used as a means to criticize the non-performing department, and perhaps an excuse for their own non-performance. The solution to the non-performance should be working collaboratively to 'fix' the end-to-end process issues. In all circumstances, the person who caused the error should correct the error.
  - work with relevant partners involved in the end-to-end business process and relevant vendors that provide process services to the organization.

Case study: Improvement of quality without SLA

A financial organization had a strong functional structure. All end-to-end business processes were divided over several departments. The customer financial statement business unit was confronted with numerous errors every print run, which was caused by another department further upstream in the process. The customer financial statement department had started with a detailed SLA approach, which did not resolve the issues. Analysis showed that there was simply no incentive for the other departments to do things correctly the first time as they knew that the customer financial statement business unit would correct the errors. It was recommended that all errors found should be corrected by the department that caused them. The error-prone departments started to realize that it was more efficient to ensure that the data entry was correct the first time, as it took significantly more time to correct the errors later.

Message: When confronted with problems with a business process, resolve the root cause rather than the symptoms.

- Provides guidance, coaching and support to process workers.
- Process improvement is a continuous activity and the Process Steward:
  - is the focal point for process improvement suggestions and proposals,
  - is the decision-maker (facilitator) of any suggested process improvements,
  - has a full understanding of all the triggers for process improvement within the organization,
  - creates an environment and the supporting mechanisms for continuous process improvement,
  - implements all process improvement changes, including people change management, training, application enhancements and
  - decides if the suggested improvement is just part of 'business as usual' or should be a new BPI project.
- Is responsible for process automation. The process steward must be involved in all automation relating to business processes. It is important to remember that many IT applications span across various processes.
- Promotes processes-thinking and action. This means creating an awareness of the importance of business processes and process-thinking within the organization. Process Stewards are the organizations process champions who continuously promote the proper use of the business processes.
- Is responsible for process performance management. This is the on-going management of the business processes within the organization and it will be discussed in more detail in Chapter 6. In summary terms, this will include:
  - establishment (negotiation) of both process and people performance targets,
  - provide guidance and support to functional line management on 'people' targets,

- ensure that targets add value to, and are consistent with, the organizational strategy and objectives,
- report and analyse the actual performance levels as compared to the established and agreed targets,
- ensure the relevant people (management and team leaders) act on this information,
- feedback and feed-forward loops are relevant and appropriate to the process and the organization and
- progressively move from a reactive to a proactive state and then on to the ultimate goal of being able to be predictive.

- co-operates with other process-related roles, such as, process modelers, process consultants, process co-ordinators, process analysts.

As stated by Air Products and Chemicals Inc. (APCI) (APQC, 2005):

> Process owners must have well-established, actively aligned, and collaborative relationships with the business units and their leaders. Process owners develop and share critical key performance indicators and targets with the business units, and they are continually identifying opportunities to improve processes across the businesses.

The required skills to be a successful process steward are described in Appendix A.

As the organization agrees upon the process stewards role and responsibilities, there may become a need in the organization for someone to be responsible for a group of processes. This person may become known as the process executive. Unlike the process steward role, this will probably initially be the one person and usually a senior person within the organization. The sophistication and acceptance of the role will evolve with the maturity of the organization.

Process Executive

The Process Steward will need to report to someone in the organization and this is usually a Process Executive. These are usually senior functional business executives within the organization who are capable of understanding and taking an organizational viewpoint rather than the narrower functional silo view for which they may be responsible.

Responsibilities

The typical responsibilities that a Process Executive will have include:

- ensuring that the processes are always in alignment with the organizations strategy and objectives. That the processes are always adding value to these objectives, and if not, and the processes are still necessary to the organization, then the Process Executive should consider the outsourcing of the process. The alignment of processes and strategy is achieved by:
  - the continual development of an appropriate process architecture.
  - understanding the impact of strategic choices on business processes. Refer to Chapter 7 for more detail.

  - co-ordinating various organizational strategies and aligning them with specific process strategies and ensuring they support the organizational goals.
- ensuring that the business processes are always customer-focused, by:
  - maintaining a constant customer viewpoint.
  - maintaining customer satisfaction and clearly understanding the difference between customer satisfaction and customer service (refer Jeston and Nelis, 2008, Chapter 17).
- obtaining the results required by the organization strategy and objectives by:
  - the business processes performing according to the established targets.
  - the Process Steward role and responsibilities being clear and concise.
- collaborating across the organization, especially functional silos, to ensure:
  - process changes are approval in a timely manner.
  - final arbitration is provided, where necessary, to ensure problems, disconnects or gaps that arise across departmental lines are resolved to the satisfaction of all stakeholders.
  - overall quality management is achieved and that quality is an integral part of every process.
- sustainability of process improvement and management is maintained and enhanced by:
  - providing support and coaching for the Process Stewards using the best resources, whether internal and/or external.
  - ensuring that Process Stewards have sufficient time to execute their role and are provided with the necessary resources.
  - ensuring that appropriate process measures/targets are established, monitored, maintained and continually optimized.
  - process models are maintained in their current state.
  - nurturing on-going and continuous improvement programmes for business processes.

Sample from APCI

APCI call this role a 'Global Process Executive' and the position accountabilities have been divided into four groupings:

Leadership:	*Design:*
Drives strategic alignment and customer focus	Defines business and customer inputs and outputs of the process
Prioritizes global improvement opportunities through annual planning process	Documents the process activities and approves changes

(Continued)

(Continued)

Resolves cross-process issues	Prioritizes enterprise process IT spending
Leads the change to a process-focused organization	Ensures controls are in place, validated and tested for accurate financial reporting (SOx)
	Audits work practice compliance
Performance:	*Improvement:*
Implements metrics and reports process performance	Analyses process performance gaps
Achieves process metrics targets and goals	Develops plans to close gaps
Prioritizes performance gaps/shares successes	Executes Continuous Improvement projects across business units
Provides adequate process resources	Benchmarks and adopts Best Practices
Monitors data quality	Fosters new Continuous Improvement ideas

Source: Air Products and Chemicals Inc. (public domain)

The division of these responsibilities/accountabilities into four different areas could be considered worthwhile within some organizations.

The required skills to be a successful process executive are described in Appendix A.

Process Executive and Process Steward selection

One of the most challenging aspects of BPM is to ensure that the accountability and responsibilities of the business processes are clearly and appropriately assigned. In the previous sections we have suggested that there should be both a Process Executive and a Process Steward. Who and how these are appointed is a significant challenge for the organization and has the potential to have a huge impact on successful outcomes.

Simplistically, 'an organization has a number of choices with regard to process ownership; the organization could:

- make the functional managers responsible for their own part of the process only (part of an end-to-end process, that is, a sub-process),
- appoint a functional manager to be the process steward and responsible for the entire end-to-end process and

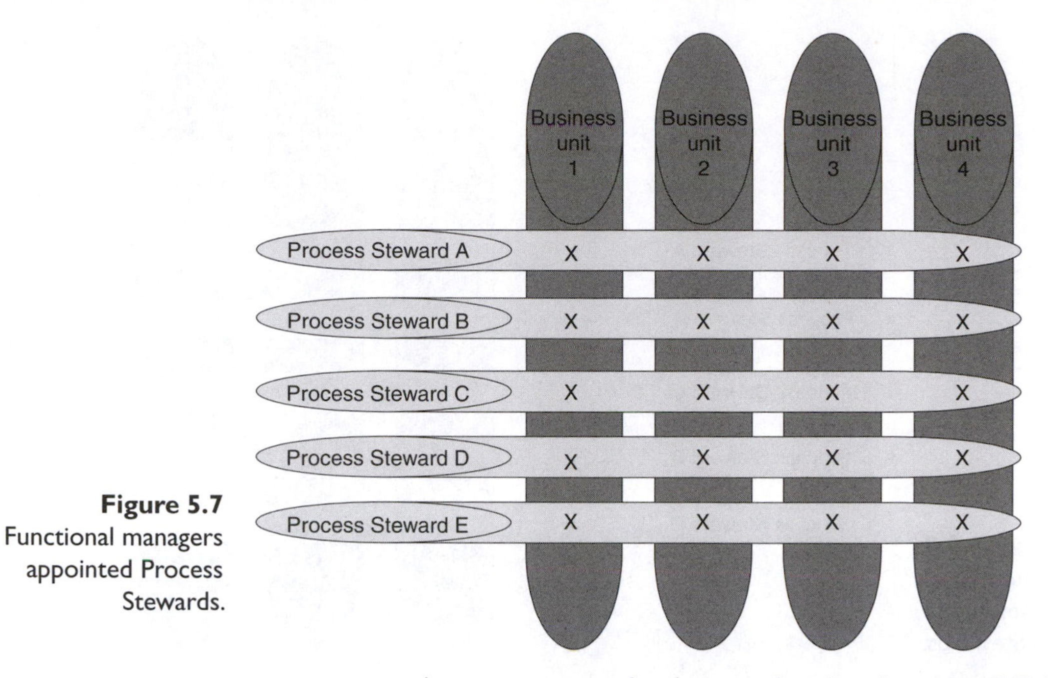

Figure 5.7 Functional managers appointed Process Stewards.

- appoint a manager who has no functional responsibilities to be responsible for the entire end-to-end process' (Jeston and Nelis, 2008, chapter 28).

We have shown in the next three figures how these choices may look and then describe the associated challenges, risks and benefits.

Figure 5.7 shows the situation where a *functional sub-process steward* is appointed. This is the situation where no one person has overall responsibility and accountability for an end-to-end process. The only *benefit* to an organization is that it is 'cheap' – that is, it will add no cost because it just becomes part of an existing person's responsibilities.

The *risk* associated with this approach is that sub-process owners will only see their own part of the process (a silo perspective) and changes to this sub-process may negatively impact other parts of the end-to-end process, which could in turn lead to a sub-optimized situation. This approach is difficult to make work in a practical sense within an organization. Process stewards meetings will be large and probably ineffective; it will be more difficult to gain consensus on process improvements and measures; the organization risks dilution of accountability and the creation of a 'blaming' attitude when performance targets are not achieved.

The second approach (Figure 5.8) is to appoint an existing *functional manager* as the end-to-end process steward.

The *difficulty* with this approach is that there is a conflict of interest. Being responsible for one or more end-to-end business process(es), and a particular functional silo (a sub-process), may lead a process steward to make decisions and changes that positively impact their own functional silo's (departmental) profitability and operational efficiency, but negatively impact the overall end-to-end process. Management of processes in this manner has the potential to lead either to the end-to-end process not being sufficiently considered or to the functional managers using their position to pursue their own functional

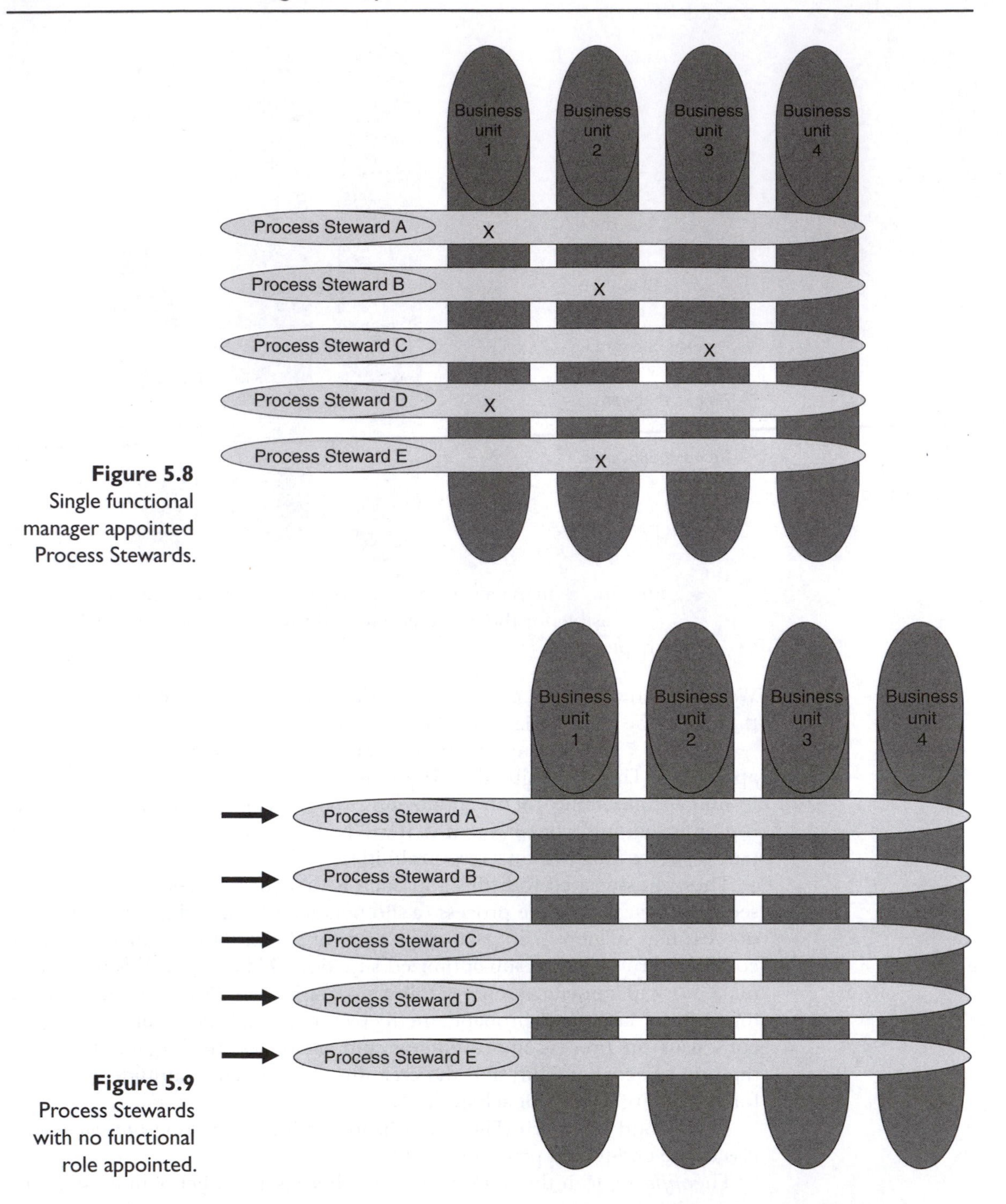

Figure 5.8 Single functional manager appointed Process Stewards.

Figure 5.9 Process Stewards with no functional role appointed.

objectives. While the approach is better that the one shown in Figure 5.7, it is not ideal.

The last suggestion, as seen in Figure 5.9, is to appoint a *process steward with no functional responsibilities*, that is, someone outside the various functional silos. The benefit of this approach is that it does not suffer from the issues associated with the first two approaches.

A drawback of this approach is that it will be more expensive for the organization and can be extremely challenging to manage because of the need to gain consensus across functional managers. For this to work effectively, the process steward appointed must be a senior executive or manager with a high level of respect within the organization and personal charisma and negotiation skills. This person must be able to provide the additional persuasion that functional managers sometimes need to look from an end-to-end process perspective. The challenge is for the process steward to be able to counter the potential sub-optimization efforts of the functional managers and pursue the end-to-end business process objectives.

From this growing maturity will come the understanding for the need of process management and control across the organization, and hence the need for a CPO and SPC (which has been described previously in the visionary section). With increasing maturity and understanding will also come increasing sophistication of the various roles and responsibilities.

Chief Process Officer

If management of an organization's processes is perceived as strategically important, then the appointment of a CPO is a significant step in the right direction and will usually follow on from the process stewards and executives. It has been suggested by some industry observers that the CPO is a role that combines the COO and the CIO. In reality, there are very few CPOs within organizations across the world. However, the appointment of a dedicated CPO is considered by many to be the ultimate way to ensure that business processes receive the maximum commitment and attention from executive management.

Role

The CPO is responsible for ensuring that the processes are geared towards contributing efficiently and effectively to the objectives of the organization. This can be achieved by ensuring that the organization's process architecture is well embedded within the overall enterprise architecture, the processes are considered with any major change or initiative within the organization, and the CBI is accepted and well respected for its contribution to the business. The role brings together a focus of the Process Executives role in a co-ordinated way.

Responsibilities

The CPO will be responsible for co-ordinating the various organizational strategies and aligning them with the specific business process strategies to ensure that they support organizational strategic objectives. This will include looking at the following aspects:

- customer service
- new product development
- procurement strategy
- fulfilment strategy
- human resource and training strategy
- accounting and finance strategy.

The CPO will be responsible for all end-to-end business processes within the organization, which might extend to the processes the organization has with its customers, suppliers and partners. This also involves the IT-related processes. As mentioned previously, IT is aimed at supporting the business processes; a separation between the two domains will lead to sub-optimization.

The Process Executives will report to the CPO from a process perspective, not necessarily from a functional perspective.

The CPO will be responsible for:

- coaching and mentoring process stewards and executives,
- creating and gaining approval of the Business Process Engagement Model (this will be described in detail in Chapter 9),
- end-to-end processes within the organization,
- achieving the process goals across the organization, and assuring the smooth flow of data, documents and information between sub-processes,
- maintaining a customer focus, constantly working to assure that processes, as a whole, function to service and satisfy the customer,
- ensuring that problems, disconnects or gaps that arise when processes cross departmental lines are resolved to the satisfaction of all stakeholders,
- planning, managing and organizing processes as a whole,
- ensuring that appropriate process measures are established, monitored and maintained,
- establishing and maintaining the BPI project implementation framework or methodology across the organization,
- nurturing on-going and continuous improvement programmes for business processes,
- smooth running of the CBI team,
- establishing and maintaining the relationships with the BPM vendors,
- on-going knowledge management and training for BPM within the organization.

Challenges

The challenges for the CPO are as follows:

- In order to obtain the buy-in from all process executives, process stewards and senior management the CPO should clearly demonstrate his or her added value, as many process executives or stewards might consider the CPO to be an unnecessary organization overhead.
- The CPO must maintain a strategic orientation and not get too involved in the day-to-day running of the CBI, as this is a completely different role that should have a dedicated manager.
- The CPO must be able to provide added value at the executive level, as all other CxOs will have bigger departments, more people and higher budgets. Thus the CPO must have the vision and capability to deliver this vision with tangible results, which will ensure that the other CxOs provide the necessary funding, resources and people to make a process-focus successful. It will be a challenge to find a person capable of successfully fulfiling this role, as the

person must have a strategic view and also be able to have a detailed understanding of the operational business processes (without going too much into detail).

It is important to include a few words of warning. A CPO role will be extremely helpful in a process-focused organization that is mature in its process-thinking and execution. If an organization is still evolving to this level of maturity, then the best alternative is to have a process-focused programme with buy-in from the CEO and other senior executives to improve the processes, and then at a later stage appoint a CPO.

Simply appointing a CPO or establishing a CBI in an organization that is not mature enough to understand or sustain them may seriously impact the added value they can bring to the organization, and could lead to difficulty in achieving the high expectations of these roles.

Having said that an organization needs to be quite mature to appoint a CPO, it is interesting that as an organization approaches the visionary state, there will be no further need for a CPO. This may seem contradictory; however, we see the CPO role as a transitional role to ensure that an organization becomes even more process-focused on its journey to the visionary state. Once end-to-end business process-thinking is engrained in the organization, the individual executives responsible for various end-to-end business processes will oversee all process-related activities.

Process management controls

As discussed in the visionary section previously, it is important that regular process-specific meetings are conducted and we have outlined some suggestions in the following sections. However, these are a part of the journey and the need for these meetings will diminish with time as management become more process-focused and understand the need to discuss and manage key business processes as part of their day-to-day activities. While we have suggested initial timings for these meetings, it is more important to ensure that the frequency of the meetings, at the various levels within the organization, is aligned.

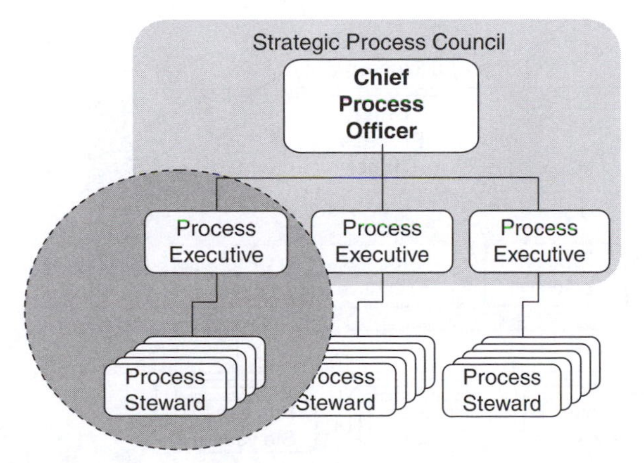

Figure 5.10 Process review and decision meeting (Process Stewards).

Process Stewards meetings

Figure 5.10 shows that the various process stewards reporting to a particular process executive should meet on a regular basis. The frequency obviously depends on the organization and issues and challenges the process stewards are facing. We would suggest that the meetings should start on a monthly basis and then could go to every two months once things are running smoothly.

The meeting would comprise the process executive and various process stewards and the manager of the CBI group. The CBI manager will provide the link to other similar groups across the organization to enable the sharing of ideas and to ensure that the groups stay on track with a common purpose.

The purpose of these meetings would be to:

- share ideas across the organization,
- resolve any issues across the business processes that the individual process stewards are responsible for. For example, a 'bordering' process steward is not working according to the agreed guidelines or in the best interests of a business process,
- manage process stewards and bordering processes. For example, on what issues should process stewards be making basic agreements, such as, quality of input, timelines and quality of hand-offs between processes and
- discuss and input into the various roles and responsibilities, operation of the business unit, process and project implementation issues, decision-making, process performance target results and appropriateness.

Process Executive meetings

Figure 5.11 shows that the process executives should meet on a regular basis with the CPO. The frequency obviously depends on the organization and issues and challenges the process executives are facing. We would suggest that the meetings should start on a monthly basis and then could go to every two or three months once things are running smoothly; however, this will depend on the purpose of the meeting and the issues being faced.

The group would not only comprise the CPO and the various process executives, but could also include the CIO and COO. The question arises, is this meeting a duplication of the SPC? It could well be, however, the purpose

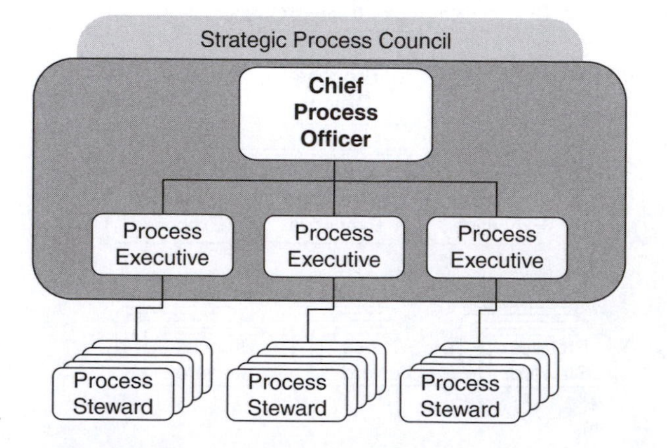

Figure 5.11 Process review and decision meeting (Process Executives).

is different and it is up to the organization to decide whether or not this is a separate meeting.

The purpose of this group is to discuss in detail the operational effectiveness and efficiency of the key or main business processes. They need to determine if the business is receiving the operational support it needs from IT and other supporting parts of the organization, and if necessary, how these support issues need to be improved. The meeting is not about investment decision-making or making decisions on the various guidelines and procedures.

Compliance and risk management

These are areas of growing importance over the last several years. The Sarbanes–Oxley Act of 2002 (SOx), also known as the *Public Company Accounting Reform and Investor Protection Act of 2002*, was almost unanimously voted into legislation by the US Congress and Senate as a result of the corporate failures of Enron, Tyco International and WorldCom, to name a few.

Organizations around the world have been required to focus on their financial processes like never before, which has meant they have been required to document them and have the CEO and CFO sign-off annually testifying to their accuracy and currency. This has meant that senior executives within organizations have had to take ownership of these processes, although few of these executives have been identified as, or called, process executives, stewards or owners.

This has provided an opportunity for organizations to build upon these documented processes and review them for process improvements.

At the very least, it has meant that when an organization improves or changes their processes, they are required to build in the SOx compliance checks and store the process models in a SOx compliant manner.

This legislation and others have meant that the attention to risk management has equally been of high importance. An organization must establish appropriate compliance and risk sign-offs into their BPI projects and process improvement activities. It should be the responsibility of the process steward to gain this sign-off and ensure that it is reviewed and checked on a periodic basis.

However, it is important to note that although regulation forces organizations to model its processes, it is important to maintain a business and customer focus while taking the regulation obligations into account.

Business case

We would have thought the need to justify a project with a business case was obvious, but unfortunately and surprisingly many organizations either do not require the completion of a business case, or accept a rudimentary or poor business case as justification for projects. This will clearly impact governance in an organization.

The purpose of a business case is to assist the investment decision-making process to allocate an organization's limited investment resources in the best manner to deliver on the organization's strategic objectives. It will also greatly assist in the project prioritization process. If an organization has no effective business case process, then they risk inappropriate allocation of these limited investment funds and inaccurate prioritization.

The remedy for the organization is to create a business case template and completion instructions, and conduct training programmes for both staff and executives. The organization then needs to establish the decision-making body for project investment decisions and prioritization. In the first instance it is unlikely to be the SPC as it may not exist in the initial process-focused journey. The establishment of the SPC will take time and maturity for the organization.

We have chosen to pay particular attention to this topic as it is an often overlooked aspect of investment decision-making, prioritization and project management. If the decision-making body understands the importance of business processes to organizational success, they are in a unique position to be able to influence and grow the process-focused journey.

Part of an organization's governance framework must be an agreed rigorous manner of evaluating suggested business investment decisions which inevitably result in projects. This may sound like an obvious guideline to have, but in many organizations projects are commenced either without a business case at all, or a very immature and inadequate business case.

How can senior executives, who are responsible for the management of an organization's finances and resources, diminish them without a thorough analysis of the best course of action and the various alternatives?

The development of a standard business case template and evaluation methodology is a critical part of the governance framework.

Writing a business case can be one of the simplest or most difficult activities one can undertake. It is arguably the most important activity in successful projects and necessary to effectively determine the appropriateness of the investment. Not only does it determine if the project is justifiable in the first place, it then guides the project during its execution and the subsequent realization of the business benefits.

For a more detailed examination of a business case, refer to Appendix B.

Excessive governance

On occasions an organization will establish too much governance. This is often caused by an over-enthusiastic drive for control. Excessive governance can actually lead to a reduced level of efficiency and effectiveness in the organization and its business processes.

Some of the ways to address excessive governance include:

- Raising general awareness around governance and explaining what is required and what is not required.
- Regular independent reviews, as the people involved in the governance are often too close to the detail to effectively challenge or review the current situation. An independent review will provide useful insights into the gaps and any excessive governance.

- Introduce a 'bureaucracy buster' who will challenge the additional effort required for governance. This can also be a role assigned to different people during workshops and meetings.
- Ensure that governance is appropriate for the size and risk of the organization and business process. A simple business process with limited or no impact on the overall organization will require a lower level of governance to the key organizational business processes that can adversely impact the entire organization and customers if ignored or mismanaged. An example is shown below:

	Low level governance	*Medium level governance*	*High level governance*
Level of effort	Less than *x* FTE involved in the process	Between *x* and *y* FTE involved in the process	More than *y* FTE involved in the process
Contribution	Contributes less than $*x* (or *x*%) profit/revenue	Contributes between $*x* (or *x*%) and $*y* (or *y*%) profit/revenue	Contributes more than $*x* (or *x*%) profit/revenue
*Business risks**	Low	Medium	High
*Impact on the organization***	Low	Medium	High
Competency and availability of people	All people are competent and available for the process	Most people are competent and available for the process	Some people are competent and available for the process

*Business risks (in case something goes wrong in the process):

- Impact on the share-price
- Impact on revenue/profit
- Impact on customers
- Impact on partners
- Impact on employees
- Government or legal implications

For each of these aspects low, medium and high criteria can be specified.

**Impact on the organization can assessed by:

- Number of departments involved in the process
- Number of people affected by the process
- Amount of management resources required

For each of these aspects low, medium and high criteria can be specified.

Key questions

If the organization is not able to clearly answer these questions (Spanyi, 2003, p. 69), then it is still on the journey towards the visionary state:

- 'Which business processes need to be improved – and by how much – in order to achieve our strategic objectives?
- What are the key measures of performance we will need to monitor?
- Which executives will be accountable for the performance of critical business processes?
- How will they be rewarded?'

Key challenge

We would like to leave you with one of the key challenges that will face an organization with regard to becoming more process-focused from a process governance perspective.

For most organizations the greatest challenge it will face in establishing a process governance structure will be the alignment with the existing functional structure. If they are not appropriately aligned and work together well, the process governance will be sub-optimal.

There is no ideal structure to resolve this dilemma. Each organization will be different and have its own way of trying to overcome this. It will need to deal with business issues, the culture of the organization and the egos of management. However, let us leave you with a thought.

If you really want to become process-focused, then perhaps the functional hierarchy needs to be subservient to the process governance structure, *from a business process decision-making perspective!*

Chapter 6

Process performance

Introduction

We stated earlier that we believe that the measurement of business processes is a significant part of the sustainability of process improvement and process management within an organization. If you link process performance to the accountability and responsibility for business processes, as we did in Chapter 5, then you have a powerful ability to actually manage your business in a high-performance management model and in a sustainable manner (Figure 6.1). This chapter is about understanding what process performance is, how it should be applied within an organization, and how it must be linked to other dimensions to achieve an organization that is managed by process and reaches the desired visionary state of a process-focused organization.

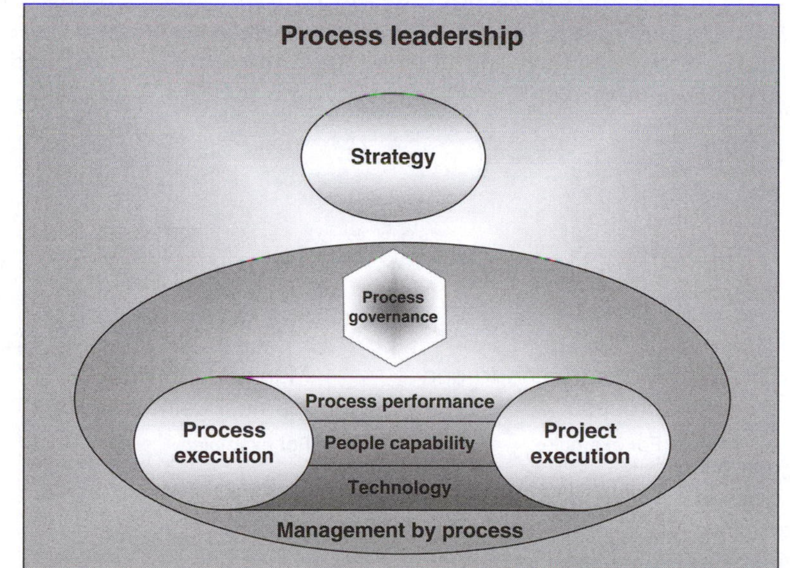

Figure 6.1 Management by Process framework: Process Performance.

Why is process performance important?

'At a time when companies in many industries offer similar products and use comparable technology, high-performance business processes are among the last remaining points of differentiation' (Davenport and Harris, 2007, p. 8). In fact, the Gartner group interviewed 1,400 chief information officers in 2006 and it was suggested that business intelligence was the number one technology priority for IT organizations (CRM, Today, February 8, 2006).

Performance establishment and measurement is about the true management of an organization's business processes in a sustainable way. There is an acknowledge truism that few people would argue with, and it is that

> if you are not *measuring* performance, you are simply not *managing* your business

and yet, few organizations effectively and meaningfully measure the performance of business processes and even fewer organizations relate rewards *clearly* to the outcome of the performance of these business processes. By effective and meaningful measurement, we are referring to the ability to immediately make decisions on how to react to a given situation, based upon the outputs of the measures. What corrective action is required? How should we do it next time? Without effective and meaningful measurement it is impossible to manage your business operations, and it is difficult, if not impossible, to continuously improve your business processes.

Before we start to discuss how and when to establish performance measurement, it is important to have a clear understanding of the distinction between what and how we are measuring, and why.

Simplistically, there are two types of performance measurements that are required in a process-focused organization as shown in Figure 6.2. The left side of the figure refers to the measurements associated with the business processes and the right side refers to the measurements associated with people (individuals), teams or parts of the organization. Both are essential and must be appropriately integrated and aligned for optimal performance.

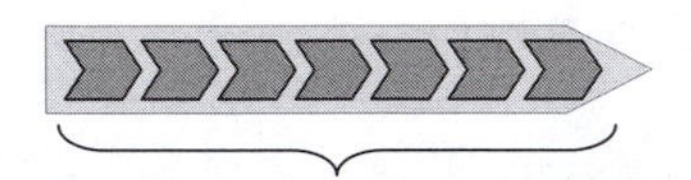

Figure 6.2 Types of performance measurement.

Key trends

In looking at the key trends we thought it important to understand the situation within most organizations. We would like to bring to your attention to some research that was completed by Kepner-Tregoe, Inc in 1995. While this research may appear to be a little old, in our experience it is still valid today within most organizations. This research comprised 1,516 completed responses (out of a sample of 4,000) from 611 supervisors/managers and 905 from workers across North America.

It is important to understand the outcomes of this research because it establishes the environment within many organizations, the difference of opinion between management and employees, and provides the basis for an understanding of the role and responsibilities of managers.

While most social scientists believe that both internal and external factors contribute to job performance, 35% of managers believed that the individual's work ethic has the greatest impact on job performance. Less than 25% of managers believed working conditions (systems, structure and business processes) have the greatest impact on job performance. *Only 12% of managers actually believed that they themselves have primary responsibility for their employee's motivation (or lack of it)*; 42% of managers ranked as the number one or two barrier to good employee job performance the employees' 'laziness or lack of motivation'.

These are 'interesting' statistics and will have a significant impact upon how managers behave in an organization. In fact, these statistics are downright frightening.

We have broken the research up into a small number of headings.

Organizational commitment

Both managers and employees agreed, out of a list of eight choices, that the first priority of the organization was *customer service* and the last was *employee morale*.

Even given this, just less than 50% of employees said they were glad to be part of the organization, whereas managers though 68% of employees were glad to be part of the organization.

Employees valued as individuals

Forty per cent of employees felt valued as an individual by the organization, whereas, only 21% of managers said employees were valued as individuals.

An example of employee value was how the organization valued employee suggestions for improvements. 21% of employees said their suggestions were taken seriously, whereas 82% of managers said they valued employee suggestions. This indicates a serious communications issue.

If this is true, then process stewards have a critical job to convince employees to make suggestions in the first place, and then communicate the outcomes of the suggestions for improvement to the employee.

In fact, 48% of employees said that changes undertaken by the organization were not undertaken for their benefit. This will have a significant impact upon business process change projects and programmes and supports our view that people change management is by far the largest component of any business process improvement (BPI) project.

Performance standards

Managers replied that nearly 33% of their organizations operated with no formal employee performance system; and 75% did not see the need for an organization-wide, systematic approach to managing employees.

Yet, 20% of managers believed that employees did not clearly understand the organizations performance standards. Both managers and employees agreed that unclear job expectations are a significant barrier to good job performance.

Twenty per cent of employees think that their manager's expectation for their teams was not fair and reasonable, and yet 79% of managers think their expectations are fair and reasonable.

Feedback, recognition and rewards

Employees are largely unmotivated, unrecognized and unrewarded. This is indicated by only 33% of employees saying that their boss knows what motivates them, and senior managers stated that less than 50% of supervisors knew what motivates their staff. Yet, *supervisors think that it is up to the employees to motivate themselves.* As one manager stated: 'I can't change people: they can only change themselves' and yet one employee stated: 'I am motivated by responsibility, credit for a good job, and, of course money'.

Even if by chance an employee becomes motivated (and most employees stated they have enough internal motivation) then only 40% said they actually receive recognition or rewards (and supervisors agree). Recognition and rewards from senior management is even rarer. Sadly, most employees had to rely on their own knowledge of how well they have done.

So how does this relate to futures trends? With the exception of more organizations implementing formal employee performance review systems, we have not seen a lot change from the outcomes of this research, which is disappointing to say the least. Lack of communication is always the number one complaint from employees and management, and yet few formal communications channels are put in place and maintained. This is not to say there are no organizations that care for and engage with its employees, there are, there simply are not enough of them.

Key elements of process performance

Managing people and performance is a discipline and skill which requires a careful, integrated approach to managing all issues and aspects – from communications to behavioral consequences, from feedback to trust.

From a process perspective we have identified the following key elements of process performance:

- Business and people process performance measures (targets) must be established, agreed, documented, communicated and implemented.
- The organization must have a clear understanding of which of the organizations' key business processes they wish to measure. It needs to have these business processes modeled (documented) and clearly defined and agreed process metrics.
- There needs to be mechanisms in place to monitor the actual performance of the key business processes, managers, staff and teams against the agreed targets.
- There needs to be recognition of good performance with matching and appropriate rewards.
- A continuous improvement programme in place to contribute towards sustainability.
- A detailed, audience specific communications programme that will be *sticky* to the recipients. That is, the audience will pay attention to it and remember, enjoy and learn from the communication.

Visionary process performance

In order to obtain a visionary or optimal business process performance state, there are a number of management requirements or guidelines that must be in place. These include following:

1. Performance measures will have be specified, communicated and be the responsibility of someone in the business. The 'someone' will be defined and agreed as part of the process governance structure. These measures must include both quantity and quality of performance measures.
2. Management will have documented models and have a clear understanding of the underlying key end-to-end business processes. The purpose of these models is to enable managers to clearly understand the effects of the decisions they make and the activities they undertake on business process performance.
3. Management will have sufficient information about the current state of the business processes. This includes process metrics and historical performance measures. This data is typically built up over time.
4. Management will have sufficient measures in place to deal with the related level of uncertainty and changes within the business.
5. If the outcomes seem to be difficult or impossible to achieve, management will escalate to higher levels of management and discuss how to proceed in the situation.

The setting of targets or establishing a measurement environment alone will not ensure the organization reaches the visionary state or progress very far on the journey to this state. The behavior of individuals and especially executives/managers within an organization is not only a reflection of the organizations culture, but significantly influenced (in fact, often driven) by the targets (Key Performance Indicators (KPIs)) set by executive leadership. Unless fully supported by an appropriate rewards system then behavioral shifts are less likely to occur. It is an interesting phenomenon that while executive leadership understands that targets (and especially financial rewards) drive executives and management behavior, yet there is a significant reluctance within most organizations to change the targets and reward systems to support the required behavior towards a process-focused performance – this will not be the case within the visionary state.

Roadmap to process performance

In this part of the chapter, we will examine performance establishment, measurement, management, and rewards.

We will outline the steps that are required to implement an effective business process performance mechanism. While these steps are primarily the responsibility of the process steward to implement, they must have the support and backing of the process executive, executive leadership or senior management for it to work in practice. The steps are shown in Figure 6.3 and will be examined individually.

Step 1: Increase business process management (BPM) awareness

This has been covered in detail in Chapter 4, Process Leadership.

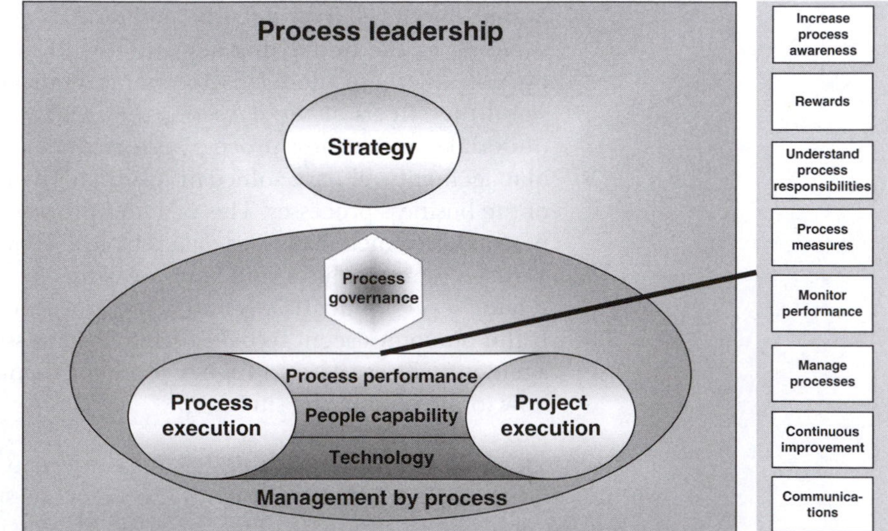

Figure 6.3 Management by process framework: Process performance steps.

Step 2: Reward determination and establishment

When an organization has *operational* challenges, whether not meeting service level agreements (SLAs), processing backlogs, lack of profitability, poor customer service, then there are broadly three areas the business should review: organizational structure, business processes and staff performance. We have always believed that the last area for management to judge is staff performance. Staff may be performing the best that can be achieved with the business processes, technology systems and organizational structure that management has provided.

The order in which to tackle operational performance issues is to:

1. review and 'fix' the business processes (this includes business process performance targets establishment, management and rewards systems);
2. review the organizational structure to support these business processes;
3. then and only then, after the first two have been well developed and fully implemented, can staff performance be reviewed and evaluated.

As part of the first two activities mentioned above, management must ensure that the motivation and rewards systems appropriately support the business process performance measurement and the organizational structure. This is further validated by Spanyi's comment: 'the details of the company's measurement and reward systems should be predicated upon an understanding of both business processes and structure' (Spanyi, 2003, p. 101).

What drives people performance within an organization is a complex set of circumstances as seen in Figure 6.4. This figure has been derived from the work of Victor Vroom (1964) on Expectancy Theory which deals with motivation and how management can influence it. While this work seems

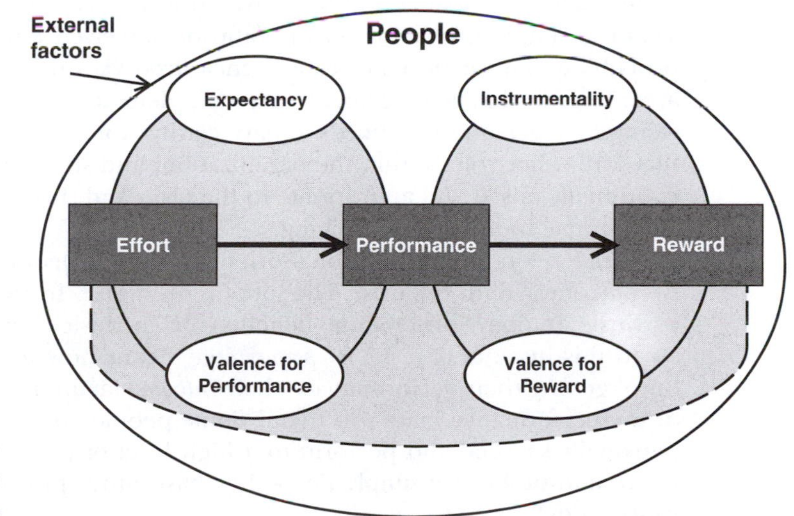

Figure 6.4 Performance determinants.

chronologically old, it is still incredibly relevant today. Vroom's theory assumes that behavior results from the conscious choices people make from the various alternatives that are available to them. Together with Edward Lawler and Lyman Porter, Vroom suggested that the relationship between people's behavior at work and their objectives or goals is not as simple as was originally imagined. Vroom realized that an employee's performance is based on individual factors such as personality, skills, knowledge, abilities and significantly their motivation for the task in hand.

Expectancy Theory states that different individuals will have different sets of goals and that they can be motivated if they believe that:

- there is a positive correlation between effort and performance (*expectancy*);
- favorable performance will *result in a reward* (*instrumentality*);
- *valence for reward* is the degree to which the person wants to earn the reward on offer;
- *valence for performance* is the degree to which the person wants to carry out (perform) the given task regardless of the reward offered.

So what does this mean in layman's terms? For an organization's board to expect management and staff to exert *effort* to deliver *performance*, then there will need to be *rewards*. While the rewards and targets (KPIs) will drive performance and outcomes, especially at the senior levels within an organization, this is not a simple task. Get the mix wrong at the peril of the organizational performance.

In order for management and staff to expend the *effort* to deliver *performance* they need to believe that there will be a reasonable expectation that *performance* will follow and be an outcome of the *effort* spent. According to Expectancy Theory there are simplistically two components that will 'motivate' management and staff to believe that performance will follow from exerting *effort* and they are *expectancy* and the *valence for performance*.

Expectancy is the estimate by people (management and staff) of the probability that performance will be delivered as a result of the *effort* expended. This is a simple probability of the ratio of perceived effort to performance probability. This expectancy can be measured via questionnaires and there are activities that management may complete to raise this *expectancy*. For example, a person's role needs to have clarity, not be ambiguous and in conflict with other roles within the organization; and skill levels may need to be continually raised and appropriate to the tasks and this can be achieved with appropriate training and coaching.

Valence refers to the emotional orientations that people hold with respect to outcomes and rewards. The depth of desire from an employee for extrinsic (money, promotion, benefits) or intrinsic (satisfaction) rewards. So in this instance, *valence for performance* means how much the people are 'into' getting their performance – their *into-ness* factor. Does the achievement of the performance matter to them? Some people are driven by the need to personally succeed and perform to a high level of performance or achievement, and others are simply not – they have other priorities in life (family, sport, study).

Once people believe that if they expend the *effort*, *performance* will follow, the next question is 'so what?' Will the effort be worth it? Even if a person has a low score on the *valence for performance* scale, they may be motivated if the reward is either high enough or something they would find highly desirable. This provides management with an opportunity of creating a clear link between *reward* and *performance*. If a clear link is not provided, motivation diminishes either immediately or certainly over time.

The first linkage between *performance* and *reward*, as seen in Figure 6.4, is *instrumentality*. While this is not a particularly 'user' friendly term, it refers to the means (desired motivation) that delivers the desired *performance* outcome. *Instrumentality*, like *expectancy*, can be measured via a questionnaire and is the probability that if the *performance* is achieved what is the likelihood of receiving the *reward*? This may seem obvious, but how many times have we all been promised rewards that never eventuate? Like the salesman who has the sales manager's role 'dangled' in front of him as a possible reward if he meets his targets (KPIs). When he achieves the target, the Chief Executive Officer (CEO) announces that his son-in-law has just been appointed as the new sales manager. Once again, management must make a clear and unequivocal, unbreakable link between *performance* and *reward*.

The second component is the *valence for reward*. That is, is the reward something that will motivate the people or individual and do they 'want' the reward? The reward must be something that the people will expend significant *effort*, to achieve the *performance* to receive the *reward*.

Rewards need to be relevant for the potential recipients and should be socialized (agreed) before hand to ensure it is something that employees find worthwhile and strive towards. For example, if the sales manager is a young single man who loves to party and he establishes a reward of a weekend away to a 'party' location, for one person; this may not be appealing to all his sales people, the majority of whom are married with children. In fact, if the sales people are 'team players', the fact that only one of them will win and receive the reward may work against the 'team' culture. The reward(s) need to be thought through and be appropriate for the people, organization and level of performance expected. Remember, rewards do not need to be monetary in nature, however they must be 'fit for purpose'. Once the rewards system is established, it will need to be extremely well communicated throughout the organization to the appropriate people.

As stated earlier, we know of proven questionnaires and toolsets that are available and will provide measures for these linkages and allow management to motivate and implement rewards systems in an appropriate manner for their staff.

Case study: Sales team were genuine team players

A wine organization had a policy of employing high-profile ex-football players as its sales representatives. These ex-players were recognized by customers and the organization found this useful in selling its products. It also had a policy that if a sales person was the worst performer for 3 months in succession they would be dismissed. As team players, the ex-footballers were comfortable in the team environment and supporting their other team members, and so were extremely uncomfortable with one of the 'team' being

(Continued)

Case study: Sales team were genuine team players (*Continued*)

dismissed. The team members always ensured that if one of them had two bad months in a row, they would *never* have the third bad month as successful members would swap sales with the underperformer to avoid the dismissal.

Message: management needs to clearly understand their staff and what motivates them.

If rewards are provided as a 'surprise' for the achievement of an excellent performance after the performance has been achieved, again, management needs to think about the reward and make it appropriate and relevant for the person. Maybe there needs to be a 'shopping basket' of possible rewards from which to choose from. We have seen people receive rewards of expensive bottles of wine and they do not drink wine. When asked why this was selected as the reward, the sales manager simply said 'it was easy'! For him maybe, but all it did is demotivate the person, rather that reward and encourage them.

Everything within the large grey circle of Figure 6.4 labeled *People* is what we have been describing to this point and relates to individuals or teams of individuals. There are, however, external factors to the *people* that will influence performance and these include, but are not limited to:

- *Products and services* – the products or services being sold or delivered can significantly influence staffs belief that no matter how much effort is exerted, performance will not follow. For example, if the product is of poor quality and does not compare well with competitor products, then it will be difficult to sell, no matter how much effort is exerted.
- *Organizational structure* – does this support the individuals? Are they able to have decisions made quickly? Does staff know who to go to for assistance or decisions?
- Does the organization's infrastructure support performance? Do the people have the right tools of trade, be they laptops, motor vehicles, printers, and so forth.
- Organizational culture will also significantly influence performance belief. Does the organization have a culture that supports performance measurement? Many do not. We have worked with organizations where the people simply refuse to be measured and management does not have the courage to change the situation.

Management has a responsibility of ensuring that employees are motivated, supported and rewarded for their performance. Management has a responsibility to shareholders and other significant stakeholders to ensure that the organization performs and meets its organizational strategic objectives and targets, and this can only be achieved if individual employees perform. After all, organizational performance is only the sum of the individual employee performances.

Step 3: Understand process responsibility

It is important for the process steward to have a clear understanding of:

- which business processes they are responsible for;
- what each of these business processes actually do in the organization, in detail;
- all stakeholder expectations.

This can be achieved by:

- Clearly documenting all roles and responsibilities. This applies not just to process stewards, but the people executing the business processes, and any other party to the process. This is part of the Process Governance described in Chapter 5.
- The business processes themselves should be documented and available, ideally via the intranet, to all interested parties.
- Clearly understood, documented, agreed and published process metrics. Examples of this would be the costs associated with the processing an individual transaction; the total cost of a business process to the organization, relevant timings, quality and any other relevant metrics.
- All stakeholders in the business process should be clearly identified and documented. If the stakeholders are not clearly identified it will make it difficult or impossible for the process steward to fulfil their expectations. Having expectations documented, agreed and communicated to all appropriate parties is mandatory. The best method for clearly identifying stakeholder expectations is the Red Wine Test outlined in Jeston and Nelis (2008) Chapter 15.

It is only with the process steward having a thorough understanding of the business processes that they are responsible for, that will enable the next steps can be completed.

Step 4: Process measures

Why?

Process measures are completed to provide a process steward with sufficient information and feedback to enable them to analyse performance and then manage their business processes. This measurement will also allow the establishment of targets for a business process or parts of a business process, and then provide the actual performance data of the processes, for comparative purposes.

How?

The first step is to understand the organization's maturity with regard to the business and staff being measured. This will determine the simplicity or complexity of the targets.

Assuming that the organization has a high level of business process maturity, the general rules for determining and creating business process and people measures are to ensure they adhere to the principal of 'SMART' goals. That is, each target must be:

S Specific and Simple
M Measurable and Meaningful
A Achievable and cover all areas
R Realistic and Responsible
T Timed and Toward what the organization is aiming to achieve.

We will take each of these and expand upon them.

Specific and simple

- Start with a few measures and increase the number as the staff and process stewards are comfortable and the need is created.
- All measures should be simple to understand and able to initiate action from the measure.
- Measures needs to be specific and relate in a meaningful way to the needs of the business. These measures could take the form of process execution times, process activity time, process workload and distribution, process state and process costs.

Measurable and meaningful

- Quality *must* always be built into the measures. When taking action as a result of the feedback always ensure that the person who made the error is the one to correct it. Direct and immediate feedback is essential for improvement.
- There needs to be a clear understanding that there may be different drivers for different parts of the organization and the targets need to reflect this.

Achievable and cover all areas

- Ensure that the targets will result in the improvement of the business processes and the performance of business staff, the individuals executing the processes and other stakeholders.
- Look to establish measures that cover customer satisfaction, effectiveness and efficiency, adaptability and quality.

Realistic and responsible

- Ensure that staff have sufficient authority to perform their tasks.
- With people targets, always establish 'stretch' targets to allow for them to grow and exceed expectations.

Timed and toward the organization's objectives

- Each target should have a particular end date.
- Each target must directly relate to the organizational strategic objectives and ideally roll up to a balanced scorecard.
- The measures selected must not be in conflict with other process targets or goals. This applies at an individual process level, departmental level and the organization's strategy or objectives.

The SMART method was introduced by Peter Drucker for checking the validity of objectives, where the objective was established as part of a Management by Objectives programme. In the 1990s, Peter Drucker put the significance of this organization management method into perspective, when he said: 'It's just another tool. It is not the great cure for management inefficiency … MBO works if you know the objectives, 90% of the time you don't.' (http://www.valuebasedmanagement.net/methods_smart_management_by_objectives.html, accessed 25 September 2007)

Certainly defining or specifying the correct measures is a critical activity that must be addressed with care. However, an organization must start somewhere and it is better to start with a small number of simple measures, and then make them more sophisticated, if need be, as management gains experience with what works for the organization and its people.

Step 5: Monitor performance

This is about the collection and presentation of the data gathered as a result of the established or implemented process and people measures. It is about the comparison of the targets against actual performance. It is the process stewards' role, working with the business and process execution staff and team leaders, to determine the best and most appropriate form of reporting this comparison. Should the comparison take the form of reports, performance enquiry screens or performance dashboards? They could take the form of notifying or alerting the process steward (or other interested stakeholders, like team leaders) of process states or bottlenecks.

The purpose of these active measures being gathered and analysed is to feed both the operational management of the business and future process refinement or improvement. Once measures are put in place, it is critical to create various information loops to enable action. These information loops should take the form of feedback loops, feed-forward loops and ultimately predictive loops.

> The global process management teams assess processes using specific performance measures and targets. The key performance indicators of a process are leading indicators and predict performance. If a team measures increases in new customer signings, then it can predict revenue generation six months from now. If signings are going down today, that predictive process measure alerts the team to take action now to prevent future revenue decrease.
>
> (APQC, 2005)

Feedback loops

In *feedback* (Figure 6.5), the actual operational results of the measurement process will be compared with the process targets or objectives. This should provide an understanding of how the process needs to be adjusted to ensure that the next time it is executed, the process is better geared towards meeting the processing targets or objectives. This may require the adjustment of the process itself, or it could require the adjustment of resources available to the process – for example, there may be a requirement for more people to be available or it may require that transactions are routed to more or less skilled people.

The advantage of a feedback loop is that it can accurately measure the extent to which the process is meeting the targets or objectives. The disadvantage is that it only provides information after the process has been completed, and this may be too late to meet the process targets and objectives.

Feed-forward loops

In *feed-forward* (Figure 6.5), prior to the processes commencing their execution cycle, relevant influences and factors should be available to allow management to anticipate the impact of the processes and enable appropriate action to be taken (e.g. if the volume is higher than anticipated, management will need to bring 'on-line' more people for process execution). It is important to understand and anticipate the impact that the feed-forward can have on the organization's ability to reach its objectives – for example, the introduction of a new product could lead to more questions in the call center, which could impact the call center's ability to meet the objectives specified for response time.

The advantage of a feed-forward loop is that it anticipates new situations. The disadvantage is that it is difficult to obtain all the necessary information and then determine the impact on the business processes.

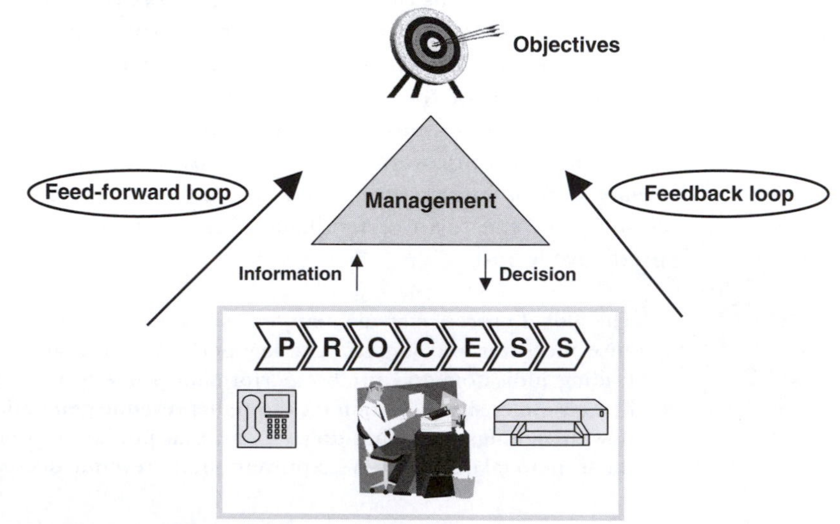

Figure 6.5
Feed-forward and feedback loops.

It is clearly better to combine all forms of information loops to enable anticipation, monitoring, managing and correction. It is crucial that the feedback loop information allows management to consider not only the process-related issues, but also the measures that have been put in place to monitor the loops (Jeston and Nelis, 2008). This will provide better insight into how the process is being impacted by changes in circumstances and/or the related management measures.

The data from these information loops needs to be channeled to the appropriate people within the organization. The process steward(s) is one of these people, but not the only person. The people who can remedy or modify the situation should be the recipients.

Once the feed-forward mechanisms have been used for some time and refined, they will lead to the business being able to *predict* the impact of certain data upon processing capability.

Predictive loops

While *predictive* loops are the most sophisticated and difficult to create, they are also the most valuable. This is the ability to predict the future situation. For example, there is a call center in London that can predict how much busy there will be later in the day based upon the call patterns of the first hour. This data has been built up over years of study and analysis, however, if the pattern shows that the day is going to be exceptionally busy, the call center is able to bring in additional part-time staff for the predicted busy times, which are usually the lunch time from 12:00 noon to 2:30 pm.

This is often referred to as part of *analytics*. Davenport and Harris (2007, p. 7) define analytics as 'the extensive use of data, statistics and quantitative analysis, explanatory and predictive models, and fact-based management to drive decisions and actions… Analytics are a subset of what has come to be called *business intelligence*.' They go on to say that 'what's left as a basis for competition is to execute your business with maximum efficiency and effectiveness, and to make the smartest business decisions possible. And *analytical competitors wring every last drop of value from business processes and key decisions.* Analytics *(which includes predictive loops)* can support almost any business process' (Davenport and Harris, 2007, p. 9).

Step 6: Manage processes

This is where the information gained as a result of the previous step is used to manage the business. It is where the daily business process performance is optimized both at an individual process, group of processes and personnel performance level. *It is where measurement meets management action.*

This management of business processes covers the implementation of the daily operational management needs of the business and its processes.

The process steward may choose to not only compare the actual performance to the established targets, but also make comparisons against other benchmarks. These could be: other business units within the organization, other organizations within the same industry (competitors), or other appropriate industries.

If benchmarking is to be used, then it is critical that the comparison is valid. It could include profitability, costs, quality, customer service and satisfaction, cycle times, and processing times. Also, benchmarking could be conducted at different levels, such as a product, process, business unit or organizational level.

Step 7: Continuous improvement

A business is not a static environment, it is dynamic and thus business processes need to be continually adapted, managed and improved for an organization to stay competitive. Hence the need for the continual improvement of the business processes to meet the changing business requirements.

While the data information loops referred to previously is an important part of this process, it is equally, or more important, to create a cultural environment where personnel are comfortable with, and rewarded for, providing proactive comments upon business process efficiency and effectiveness, and how the processes may be further improved. This culture should be encouraged until it becomes 'just what we do around here'. This cultural state may be encouraged by establishing a reward structure that provides benefits to staff for innovative, quality and efficiency suggestions.

If the suggested process improvement are minor in nature, then it is the responsibility of the process steward to manage the implementation. Where the suggested process improvement is large enough, the process steward should commence a new process improvement project. The process governance framework should define where a minor improvement becomes a project.

Step 8: Communications

Communication is the hub of all these activities. The roles and responsibilities, process documentation, targets (process and people), gathered and analysed data, action taken, must be communicated to all stakeholders in a simple manner that can be easily understood, remembered and acted upon.

Negotiation, collaboration, compromise and agreement are all critical aspects of making this work. It is essential that this is not solely viewed at a management level. The staff *must* be part of the process and are critical in an effective communications strategy.

Once again the organizations' current starting point will determine the level and sophistication of the measures and resulting actions (management).

Surprisingly, most organizations have poor or no real written measures for staff. Many organizations have attempted to implement a Balanced Scorecard and have KRAs (Key Result Areas) or KPIs for individuals, but few measures relate to business processes and even fewer allow for the genuine management of the business processes. Again, few organizations have much more than rudimentary business process performance measures above SLAs. One of the issues with the Balanced Scorecard approach is that it does not make a clear explicit relationship between business process performance quadrant and the customer requirements quadrant, however, Strategy Maps does address this. This, however, does not mean an organization should avoid using the Balanced Scorecard approach, it simply means that the organization must make the link itself and explicitly communicate it throughout the organization, especially to those responsible for delivering on the objectives.

Figure 6.6 shows the journey for an organization that starts with an unwillingness (or inability) to performance measure its processes and staff, to the state of a mature process-focused organization.

Stage 1 introduces some initial simple business process measures which will be monitored to provide feedback to allow for the effective administration of the process(es).

Stage 2 will be embarked upon once the management and staff is comfortable with the journey. It shows that more realistic process measures have been agreed and implemented. This provides management with information to allow for the 'management' of the processes. It will incorporate feedback loops that allow for the 'reactive' management of the business processes.

Stage 3 initiates stretch targets for the business processes and the initial introduction of measures that relate to processing people (staff). It will also have enabled historical data and models to be accumulated that will allow the introduction of feed-forward loops and the resulting ability to proactively management the business processes. Continuous improvement will also be activated, because the appointment of process stewards should have also occurred by this stage.

Stage 4 and beyond is about making the measures or targets more realistic for the performance of the people, and having quite sophisticated business process measures. Additional and more detailed data will progress towards

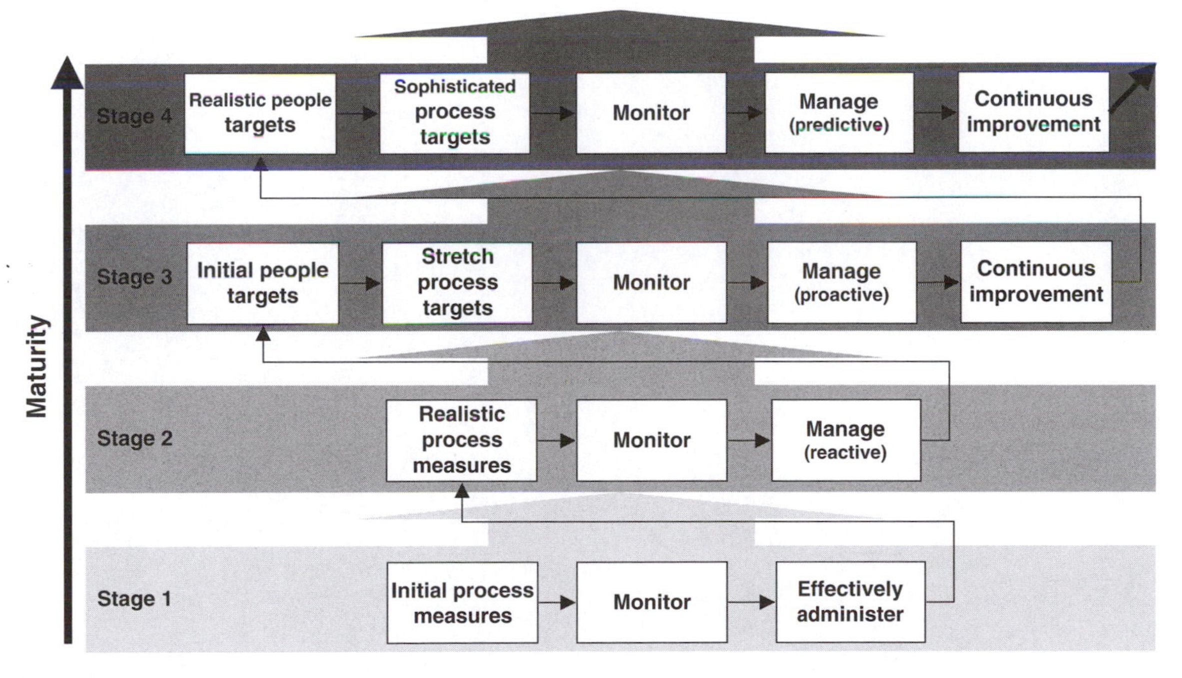

Figure 6.6
Process performance at different stages of maturity.

having the ability to be predictive about future performance, and of course, improvement will be ingrained into the culture of the organization and its people.

There is no point in implementing sophisticated and complex targets for people and processes from a low base, they need to be kept simple in the beginning. The length of time it will take an organization to mature and absorb performance measurement will depend upon both the type and maturity of the corporate culture, management and the individuals within it. People change management is a critical activity with the introduction of performance measurement and people must be made to feel part of the journey.

When auditing process-focused initiatives we have often found that the employees are enticed to participate by their dedication for either the organization or the BPI philosophy. Unfortunately, for many of these employees their current KPIs are based on quantitative targets not directly relating to the new business process initiatives. So these people are not encouraged to contribute.

Furthermore, the people who can really make a difference, the business managers, have in most cases no incentive at all to support a process-focused initiative as it distracts them from their short-term goals and objectives (KPIs), unless, of course, their targets are modified. This clearly shows the unequal balance and the struggle for process *passionate* people to truly make a difference in the business.

Unless the executive and senior management (leadership) get the KPIs right, they will place the entire process-focused activities at risk – possibly terminal risk.

Chapter 7

Strategic alignment

We had improvement programs, but the real difference came when we decided it was no longer a program, it was a business strategy.

(Stephen Schwarts (IBM))

Introduction

Firstly we should describe what we mean by strategic alignment. Spanyi (2003, pp. 97–98) said it succinctly when he described 'organizational alignment as the degree of 'fit' between an organization's strategic direction, its business processes, its structure, performance measures and rewards'. Simply, this is what this chapter is about. We will examine the strategic alignment (strategic direction) dimension that aligns process execution and management with the strategic objectives of the organization (Figure. 7.1).

Without this alignment the business processes will inevitably deviate from the key drivers and goals of the organization, resulting in organization wide frustration: employees will lack the required guidance, management feels powerless as their plans are not executed correctly and finally the customers will be disappointed as they will not receive the value they expect.

This chapter transcends the alignment of individual projects or individual processes with the strategy and it relates to the systematic and continuous alignment between the strategy and objectives of the organization with the business processes execution and management. This cannot be delegated to an individual project manager or lower-level management.

Process improvement projects are often started by linking it to the strategy. However, during the design and implementation phases of the project the alignment with the strategy starts fading away. At the time of embedding the project into the organization and making it part of 'business as usual' (BAU) process operational execution the alignment with the strategy is

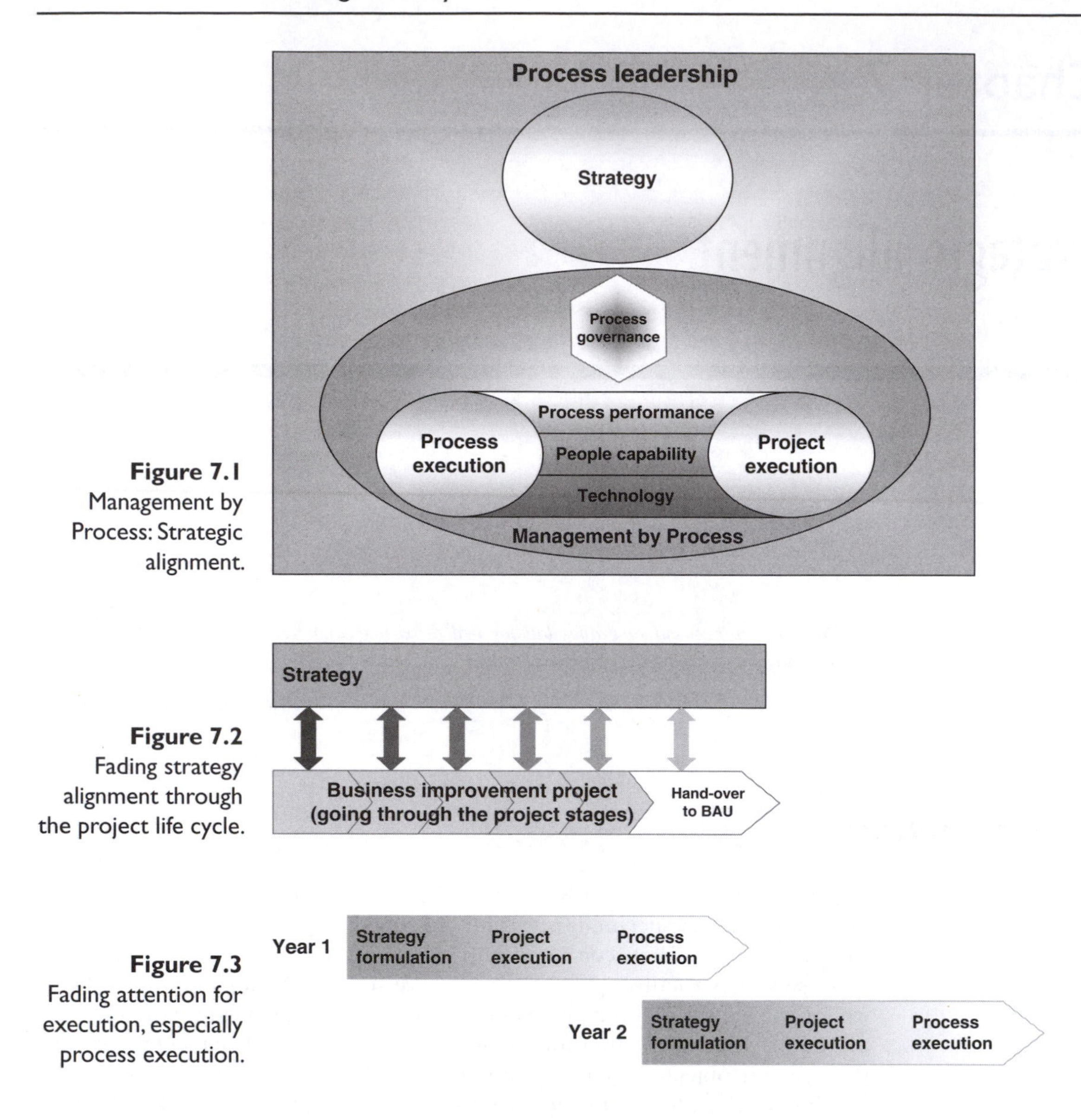

Figure 7.1 Management by Process: Strategic alignment.

Figure 7.2 Fading strategy alignment through the project life cycle.

Figure 7.3 Fading attention for execution, especially process execution.

forgotten. This can result in less than expected benefits to the organization. This is shown in Figure 7.2.

The same situation can occur with the formulation of the strategy. At the beginning much effort is put into the formulation of the strategy and the formulation of the projects to achieve the objectives. However, once the projects are up and running and have either provided some initial results, failed or are delayed, the attention starts to fade by the time it is handed over to BAU. Many organizations are tempted to go to the formulation of a new strategy rather than evaluate and correct the execution of the previous one. This is shown in Figure. 7.3.

This chapter will not discuss the formulation of an organizations strategy itself, as this falls outside the scope of this book. However, it does focus on the alignment of the formulated strategy with process execution.

Why is strategic alignment important?

We have previously defined business process management (BPM) as 'the achievement of an organization's objectives through the improvement, management and control of essential business processes' (Jeston and Nelis, 2008). Thus business processes are a critical mechanism for the achievement of an organization's strategy. Business processes need strategy to provide guidance to ensure that the right things are being performed.

The following data supports the importance of strategic alignment (Palladium, 2007):

- 95% of the workforce does not understand the strategy.
- 90% of organizations fail to execute on sound strategies.
- 85% of executive teams spend 1 hour or less per month in discussing strategy.
- 70% of organizations do not link middle management incentives to strategy.
- 60% of organizations do not link budgets to strategy.

This data supports our own experience that most organizations go through the motions of developing a strategy, but that there is little or no systematic or continuous execution of the strategy, especially as it relates to the functioning of business processes within an organization.

> We've all met managers who try to keep their employees focused on the work and let some strategic planning group think about the external, competitive world. But the attention psychology literature suggests that employees will pay more attention to their work if they understand it in the context of the competitive world.
>
> (Davenport and Beck, 2001, p. 19)

Kaplan and Norton specify in their book Strategy Maps (2004) that breakthrough results require the following three components:

1. Describe the strategy (strategy map)
2. Measure the strategy (balanced scorecard)
3. Manage the strategy (strategy-focused organization).

Oil tanker syndrome

Most executive managers are frustrated with the lack of agility and maneuverability of their organization. Much time is spent on creating the new course (strategy) and when the captain sets the new course ... it is often thought that it takes too long for the new course to take effect. As a consequence, many executives grow rapidly impatient and want to change the course again, probably more radically to make it effective. These new changes will again take considerable time to take effect. Furthermore, it is important to remember that too many changes will put too much strain on the ship (organization).

(*Continued*)

Oil tanker syndrome (*Continued*)

The short cycles of financial reporting force executives to show improved financial results for each reporting cycle, leaving little time for the necessary restructuring and changes, so that the ship can be more agile and maneuverable.

Message: It is more important to ensure the maneuverability and strength of the ship, than to keep changing the course. Stakeholders do understand that investment in the organization is required and takes time to be effective as long as the benefits of such an exercise warrant the cost and that they are delivered in a timely and managed way.

We recommend that:

- The alignment of strategy and business processes needs to be a *continuous* activity to ensure that it is actually embedded in the organizational way of working – '*the way we do things around here.*'
- The alignment needs to be *systemic*, to ensure that it is completed in a uniform, predictable, repeatable, sustainable and logical way, thus avoiding the situation where strategic alignment is the reflection of an individual's views and experience.

Process and organizational governance will greatly assist with this.

Key trends on strategic alignment

We have noticed the following trends:

- Organizations are re-assessing and re-adjusting their strategic position and their strategy on a more frequent basis which is partially fueled by the ever increasing competition. Hence, the strategy formulation and execution is becoming more and more a process.
- Middle management are becoming more aware of, and better skilled in, strategy management (through internal programmes, MBA programmes and all the available strategy literature).
- Organizational strategy has proven to be an excellent basis for the new breed of business process improvement initiatives. These initiatives are progressively spanning the organization's silo's, crossing the organizations boundaries, starting to relate to the end-to-end business processes and becoming more customer-focused. The formulation of measurable objectives provides a good starting point for process execution.
- More and more organizations are commencing business process improvement initiatives as part of their execution of strategy. These initiatives have high-level commitment within the organizations and, provided the execution is completed correctly, can provide significant and sustainable benefits. These projects have generally better governance and monitoring mechanism and commitment as their progress and outcomes are essential in the achievement of the strategic objectives.

- More organizations are realizing the importance of business processes and are making centralized, and in some cases top-down decisions regarding its business processes, although this has a long way to go before it is universally adopted and accepted. Organizations are applying different types of process architectures, process charters or strategy techniques. Some have a decentralized management of business processes. This is increasingly due to an explicit choice, rather then isolated initiatives.
- Enterprise architecture is becoming accepted as a mechanism for recording strategic choices and their consequences.

Key elements

We will now discuss the key elements of a continuous and systematic strategic alignment process (Figure. 7.4) for the key elements of:

1. the process of strategic alignment,
2. the strategic choices,
3. documenting the strategic choices and the consequences for a process architecture (in the context of an enterprise architecture).

The organization strategy and process architecture have already been discussed in our previous book (Jeston and Nelis, 2008) in the context of an individual BPM project and are considered foundations for such projects, with the type of project determining the extent to which both will be considered.

In the remainder of this chapter we will review the strategic alignment and enterprise architecture from an organization perspective rather than just an individual project.

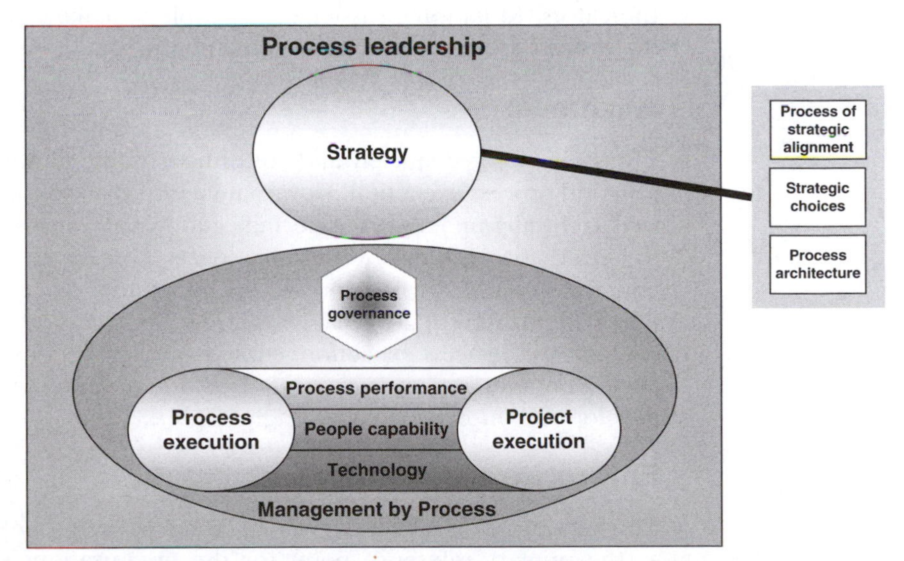

Figure 7.4 Management by Proce ss: Strategic alignment steps.

Visionary strategic alignment

Each of the key elements will now be examined in the context of visionary strategic alignment.

Process of strategic alignment

Visionary strategic alignment will create both a structure and set of standards and guidelines to move the organization forward, which will result in the creation of a sustainable competitive advantage. The visionary situation (strategic objectives) is created to provide a goal for the organization to aim towards. All business processes and their performance will be linked to these strategic objectives. Executive managers will frequently monitor the progress of realizing the strategic objectives and critically manage those that are not performing and then drill down on the underlying projects and process execution that may be causing the poor performance.

All newly initiated projects are aligned with the strategy throughout the life cycle of the project. Business owners and process stewards are project sponsors or members of the steering committee of these projects.

A visionary organization will use the Strategy Map and Balanced Scorecard approach (or equivalent) for providing the best guarantee that the strategic process is being followed and that the performance and improvement of the processes are monitored by strategic management.

Strategic choices

Organizations that have made clear strategic choices will ensure that they are able to be operationalized. The consequences of the strategic choices are also clearly specified and communicated, so they can be implemented. This information will allow staff and management involved in process execution and management to identify any gaps.

Changes to the strategic choices will be assessed as to their impact on the organization, including people and processes. Actions and Key Performance Indicators (KPIs) will be assigned to people to make the necessary changes in the process execution and management.

Alignment

The strategy is systemically and continuously aligned with the project execution and process execution. Governance is in place to monitor the strategy as well as the alignment with project execution and process execution.

Governance will have different aspects to it for the governance of strategy, the governance of process execution and the governance of project execution. Different governance will also be necessary for the links between strategy and process execution, strategy and project execution, and process execution and project execution. The organization will recognize this and develop and implement accordingly (Figure. 7.5).

Enterprise architecture

The enterprise architecture will be used by everyone in the organization as the primary reference point for the organizations strategy and the key

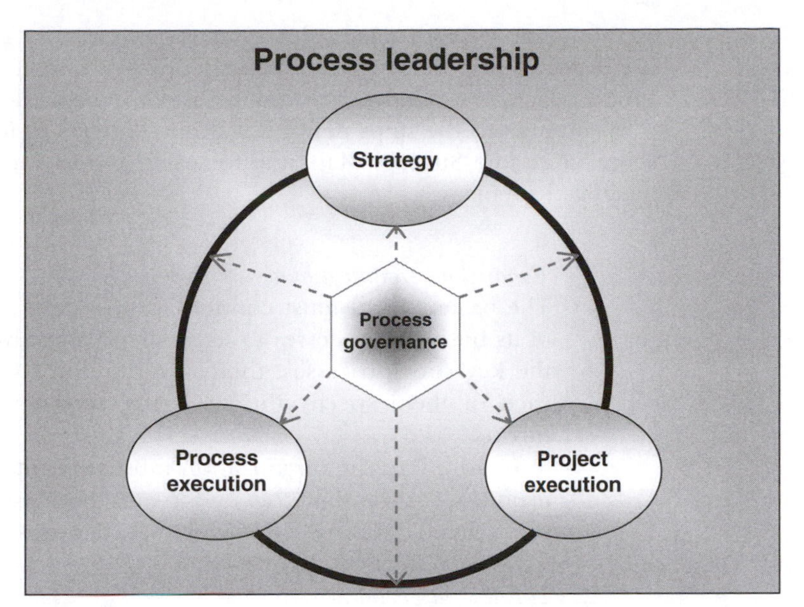

Figure 7.5 Governance and its relationship to the strategy cycle.

principles of the organization. It will also show how the alignment between the business, process and IT domain will be managed.

Roadmap to strategic alignment

Introduction

Up to this point we have described the visionary strategic alignment state that an organization could aspire towards. The reality is that few organizations are this mature when it comes to being structured and continuously aligned with the organizations strategy, process execution and project execution. Organizations will vary considerably as to where they are on the journey. Most organizations do, however, understand that 'strategic alignment' is a critical part of what they should be doing (particularly at executive level). So how does an organization get from where it is today, towards the ideal? We have included several suggestions on the impact of various strategic choices on the business processes; the positioning of process architecture within an enterprise architecture and the impact upon business processes; we have then provided several starting positions that an organization may find itself in.

Strategy Map and Balanced Scorecard with process emphasis

The Strategy Map and Balanced Scorecard approach are closely related. The Balanced Scorecard has been evolving from a performance management tool to a comprehensive strategy management tool. While the Strategy Maps are a method of providing a macro view of an organization's strategy, and provides a language in which to describe the strategy, prior to constructing metrics for evaluating strategy performance (www.wikipedia.com, accessed 10 October 2007).

We have used the steps defined by Rohm (2005) to formulate a Balanced Scorecard and a Strategy Map, and for each step we have added the process specific elements:

1. *Conduct an organizational assessment*
   The organization must conduct an assessment of the contribution of its business processes to its strategic objectives. It must highlight the key core processes, their contribution to the strategy and for each of them specify the strengths, weakness, opportunities and threats.
   In addition, the organization must appoint a cross-departmental team to ensure that the end-to-end business processes are considered and assessed. Each team member must have a strong understanding of business processes.
2. *Define strategic themes*
   Strategic themes are a method for grouping strategic objectives. The visionary organization should include a strategic theme that will target the organization to become a more agile and responsive organization. This theme should be used for initiatives that relate to continuous business process improvements, streamlining operations and enhancing customer-focus process and process-focus.
3. *Choosing perspectives and developing objectives*
   The organization should develop its strategic objectives. Examples of process-focused strategic objectives could include:
   - *Learning and growth*
     - continuous improvement will be embedded in all the work
     - cultural change to foster process-focused thinking and working
   - *Internal*
     - improve the efficiency and effectiveness of the organization
     - seamless end-to-end business processes to provide the required products and services
   - *Customer*
     - improve the customer end-to-end business process
     - improve customer satisfaction
   - *Financial*
     - optimize costs.

   Some organizations refer to these as strategic imperatives.
4. *Develop a strategy map of the organization*
   As part of the preparation of the strategy map the various objectives will be connected as part of the cause-effect linkage. These are completed from a bottom-up perspective – in other words from the learning and growth perspective to the internal perspective to the customer perspective to the financial perspective. It is often found that process-focused objectives, as defined in Step 3 (especially the

ones in learning and growth perspective and internal perspective) are instrumental in achieving the objectives of the customer perspectives and financial perspectives.

5 *Define performance measures*
When defining performance measures the organization must recognize the importance of having metrics and reporting that will adequately measure the progress and achievement of the specified objectives. Furthermore, it is important that the measures can be easily obtained and reported upon.

Details on process performance measures and establishment have been described in Chapter 6.

6 *Map to business processes (new step that we have introduced)*
The organization will add to its Balanced Scorecard a mapping of the key business processes to the strategic objectives and measures. This has the following benefits:

- Mapping the objectives to business processes rather than departments increases the process-focus and end-to-end business process thinking.
- The measures will be better aligned with the business processes and their outcomes.
- Projects will be closely linked to the business processes and the 'BAU' situation.
- Clear process ownership.

7 *Developing projects*
Projects (also called initiatives) should be developed for all the measures that *cannot be achieved* through the normal BAU activities. It is important that these projects are developed in such a way that they can be easily embedded later in the business processes – either existing or future.

8 *Computerizing and communicating performance information*
Ideally, the organization will have developed an automated solution for obtaining the performance information and it is strongly recommended to obtain the information from the relevant Business Process Management Systems (see Appendix D).

9 *Cascading the Balanced Scorecard throughout the organization*
The organization must ensure that every entity and preferably every individual performance indicator relates to the Balanced Scorecard. This is a critical step that is often overlooked. The only mechanism to ensure that everyone is focused on the organization's strategic objectives is to ensure that their performance is reviewed on that basis. This means that individual and group targets (KPIs) and actual performances are all interrelated and monitored continually. If this step is not taken then it risks people only paying lip service to the strategic objectives. This step will ensure that each individual performance is linked to core business processes and their performance.

10 *Using Balanced Scorecard information to evaluate and improve performance*
The completion of this step will close the loop of the Plan–Do–Check–Act cycle of Deming (Walton, 1986). When measuring

performance it is important to use both feed-forward and feedback loops. Refer to Chapter 6 for a detailed explanation of these, together with predictive indicators.

11 It is important that the information obtained via these methods are used to check if the stated objectives are being met and can still be met. If this is not the case, then there are basically two options: either modify the activities to improve the chances of meeting the objectives, or modify the objectives. Often organizations who have just introduced the Balanced Scorecard and Strategy Map approach establish over-ambitious objectives without taking into account the organizations' level of strategic maturity and its available capacity and capability.

Case study: Consequence of choosing the wrong performance indicator

Call centre management wished to align work effort with their stated performance indicators. Unfortunately, the organization had not chosen its indicators very wisely. The objective was to achieve a higher level of customer interaction. Management decided to reward the call centre sales staff, not just on actual sales achieved, but also on the duration of the call.

Staff admitted that they purposefully kept the customers on the telephone longer so that they would be better rewarded. In fact, they kept customers on the telephone even if the customers became irritated.

Message: Management must think very carefully and thoroughly about the performance targets that they set, because it will create behavior and it may not be the desired behavior. '*You get what you set and reward*'.

Strategic choices

Many organizations have a formulated strategy, however only a few have the execution and management of their business processes systemically aligned with the strategy – some make an effort but are not consistent throughout. A change in organizational strategy will often require at least some modification to the business processes and this is often ignored. The following types of strategies should be understood by management and staff, a distinction made between them and an understanding of the impact upon the business processes:

Operational excellence ('best total cost')

Characteristics of this strategy are lowest price, higher delivery assurance and generally limited product or service choice.

The impact on the business processes include:

- The *business processes* are tightly executed with little variation for individual customers or market segments. Modifications to the products and services are fully scripted within the business process itself – for example, Dell Computers runs an operational excellence strategy where a computer-based product configuration application allows the customer to completely modify their requirements (within set boundaries). Continuous improvement is specifically focused on making the business process more efficient and effective. Any errors need to be addressed as rework and this must be kept low as typically operational excellence organizations operate on low margins.

- Most of the business processes will be *automated* as much as possible. In most cases, the automation will extend into the full supply chain and involve partners, distributors and suppliers. Customer self-service is geared towards lowering the internal business processing costs, although it might be positioned as customer service from a marketing perspective.
- Organizational *culture* will be focused on business efficiency by improving output and reducing costs and waste. The training will be focused on the specific areas of expertise.

Product leadership ('best product')

Characteristics of the strategy are time-to-market for product, superior brand image and technological innovation.

The impact on the business processes include:

- The *business processes* are all geared towards establishing and maintaining the product leadership. Product leadership does not just relate to the physical product itself, but all interactions with the customers to reassure them that they are still dealing with the product leader. For example, leading consultancies do not just aim to provide excellent service to their clients, but will regularly publish research papers, and organize seminars and conferences to maintain their thought leadership. The quality assurance is of the highest order and may lead to higher levels of rejections and rework, as only the best quality is acceptable.
- *Automation* will tend to be more sophisticated and be more related to the products and services rather than at the operational business processes. This automation will feed the research and development, production and maintenance processes to provide the management with an end-to-end view of the product processes.
- Organizational *culture* will be focused on innovation and creativity. Continuous improvement will focus on additional features or even new products that can be established, such as the post-it notes manufactured by 3M. This requires staff to have the ability to reflect and/or experiment to extend the product features. At any given time there may be many projects related to product enhancements or new products. Staff will be trained in the products features.

Customer intimacy ('best total solution').

Characteristics of the strategy are: a high level of customization, one-to-one marketing and partnerships.

The impact on the business processes include:

- The *business processes* are quite flexible to ensure that customization is available for each individual customer or customer group. Decision-making must be delegated close to the customer so that specific circumstances may be fully taken into account. The accountability and information management need to be sufficient to be able to supervise the decisions made.
- *Automation* will be focused on obtaining and using customer information. Business processes are geared towards using the customer information to modify the business processes, products and services to meet the specific, and often changing, customer requirements.
- Organizational *culture*: Everything must to be focused on the customer experience and requirements. Training of staff will be focused on self-empowerment and delegation of authority.

Fast adaptor ('me too')

Characteristics of the strategy are assessment of the success of other innovators and product leaders, and agility to follow any innovative ideas that are obtaining market acceptance.

The impact on the business processes are

- The *Business processes* are generally well managed as they require modifications at short notice. The product development and fulfilment processes are agile, enabling quick modifications. Typically these processes have been assessed on the basis of various what-if analyses and relevant modification made.
- *Automation* will enable processes to be swiftly modified, especially to the production and delivery processes.
- Organizational *culture*: the organization has to deploy a 'can do' attitude.

Innovation ('best innovation')

Innovation aimed at creating new and uncontested markets (e.g. Apples iPod). This strategy, as outlined in '*Blue Ocean Strategy*' (Kim and Mauborgne, 2005), has a more drastic and fundamental view on innovation than outlined in the section Product Leadership strategy.

The main differences are, instead of:

- Competing in existing markets – create new and uncontested markets with either new products or new applications of existing products.
- Trying to beat the competition – make the competition irrelevant by having a unique proposition.
- Exploiting existing demand – create and capture new demand.
- Making the cost-value trade-off – break the cost-value trade-off by repositioning product and services.
- Aligning the whole system of an organization's activity with its strategic choice of differentiation or low costs.

The Blue Ocean Strategy is not just a different strategic choice, it is a paradigm shift that requires a fundamental shift in the way an organization works. Furthermore, many of the organizations that have successfully deployed blue ocean strategy will try to continue their advantage through continuous innovation.

- *Business processes* – may need to be revamped after the move towards a 'blue ocean' to ensure that they are aligned with new product proposition.
- Organizational *culture* – people in the organization are proud of their innovative work and understand the importance of continual innovation.

Enterprise architecture

The best definition of enterprise architecture that we have seen is by Wagter (2005) where he states that it is a 'consistent set of rules and models that guides the design and implementation of processes, organizational structures, information, applications and the technical infrastructure within an organization'.

Process architecture is ideally a subset of an all encompassing enterprise architecture, which will include individual architectures for business processes, the business, information and technology:

We have listed below the attributes that comprise an excellent process architecture (Wagter, 2005):

- a set of rules, principles and models for the business processes
- a basis for the design and performance of the business processes
- it is related to organization strategy and objectives
- aligned with the business architecture, information and technical architectures – which equates to an organization driven enterprise architecture
- be easy to understand and apply by all relevant stakeholders
- be dynamic – that is, easily adaptable to the evolving process, business and enterprise changes.

A process architecture should ideally been used in the context of an overall Enterprise Architecture as shown in Figure. 7.6. This figure shows how process architecture is the link between the organizational strategic objectives (business strategy) and IT architecture. Not only will it document the information in the process architecture information shown in middle box in Figure. 7.6, but it will also link

- the business strategy with the various products, services, business processes; the roles executing the business processes and who has ownership of the processes
- business processes with the application systems, data and screens that support them.

Van den Berg (2006) states that there are, in principle, three levels of enterprise architecture:

1 *Strategic architecture* – purpose is to provide support for decisions about 'far-reaching' enterprise-wide organizational business goals,

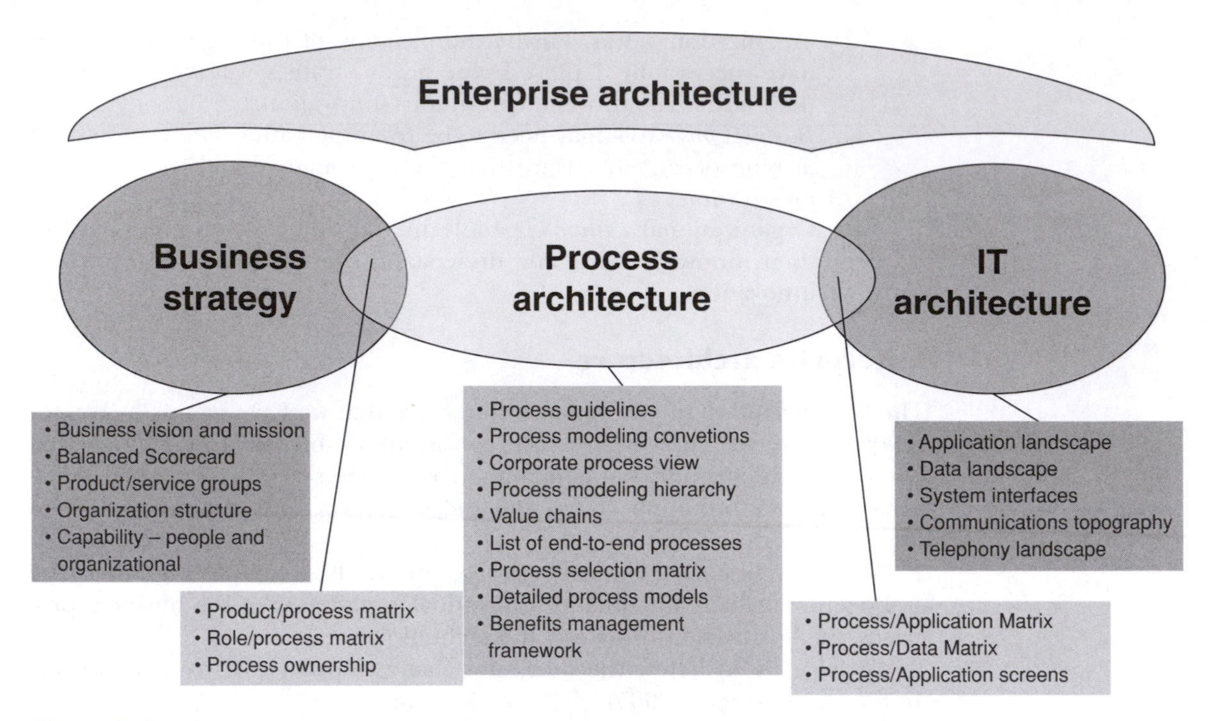

Figure 7.6
Process architecture in context of enterprise architecture.

priorities and infrastructural requirements and is geared towards senior management.

2. *Tactical architecture* – purpose is it to support decisions about the feasibility and achievability of a particular organizational goal and is geared towards middle management.
3. *Operational architecture* – purpose is to provide a concrete and goal-oriented framework for a project or programme of work and is geared towards operational management.

Table 7.1 provides a summary of the individual aspects of each of these three levels of architecture.

Development of a process architecture

The steps to developing a process architecture have been outlined in our previous book (Jeston and Nelis, 2008). The key steps outlined were:

- Obtain strategy and business information
- Obtain process guidelines and models
- Obtain relevant IT principles and models
- Consolidate and validate
- Communications
- Apply architecture
- Make it better (an architecture is never finished, it only gets better).

Table 7.1
Three levels of architecture (Van Den Berg, 2006)

	Strategic	Tactical	Operational
Purpose	Provide support for decisions about "far-reaching" enterprise-wide organizational business goals, priorities and infrastructural requirements	Support decision about the feasibility and achievability of a particular organizational goal	Provide concrete and goal-oriented framework for a project or program of work
Audience	Senior Management	Middle Management	Operational Management
Trigger	Business goals with an extensive scope (e.g. new strategy)	Business goals with limited scope (e.g. introduction of a new product on the market)	Concrete project (e.g. implementation of a new product)
Supports the production of	Strategic documents, business cases or program proposals	Business cases, project proposals	(IT) solutions
Language used	Simple	Business	Technical
Tools	PowerPoint	PowerPoint and architectural tools	Architectural tools
Focus	Coherence and collaboration	Function (what)	Design (how)
Scope	Often organizationally determined: a branch, enterprise or division	Often determined by the business goals: the functional and technical areas that are affected by the business goals	Determined by the project: the delineation of the project

While developing architecture it is important to keep in mind the current level of organizational maturity of architectural thinking and behaviour. Many architectures fail as a result of the architects overestimating the architectural maturity of the organization. For a pragmatic, yet powerful, framework on architecture maturity, see Van Den Berg (2006).

Starting positions

While we have provided a list of activities that an organization must implement if it is to move towards a visionary state for strategic alignment, we will also provide a small number of starting positions that an organization may find themselves.

No formal strategy

Situation: These organizations might have some kind of business plan, but typically lack a set of clearly defined, documented and communicated strategic choices and objectives. As a consequence the organization will struggle to achieve its targets because there is no clear guidance. Initiatives that are being started face a common problem: to obtain consensus because there is no common and unambiguous foundation for decisions.

Recommendations: It is important to obtain the key strategic choices from the relevant senior executives. It is often very difficult to obtain this type of information from an organization that has not yet specified any formal strategy. Tom Davenport (2001) stated very clearly that attention is an increasingly scarce resource. Thus, it is more difficult and yet crucial to focus the attention of the relevant stakeholders to obtain the necessary information. In other words, if a business unit wants to improve the performance of its business processes, it is better to focus on the strategic choices of this particular business unit, rather than to try and focus on the entire organization.

Six key questions to ask:

How important is project and process execution in the overall scheme of things?
It is important that an executive truly supports the initiatives. Just paying lip service is not good enough. If the executive does not include the initiative in their top five projects or initiatives, there is absolutely no change of succeeding, especially when you consider that a business process improvement initiative and the subsequent embedding in the process execution requires much more support than the occasional tap on the shoulder or an infrequent email.

Why is there no formal strategy in place?
The answer provides an indication as to whether executive management is not able or willing to develop a formal strategy. The answer to this question may provide ways of convincing management to start formalizing a strategy. Common arguments against having a strategy are:

- we do not have time – executive workshop may provide key strategic elements
- we are dependent on others – involve key stakeholders in the discussions
- everyone knows what we are doing – show that there are discrepancies and clarity is required. If it is not documented and distributed, then 'everyone does not know what you are doing'.

What are the formal key objectives that need to be achieved in the coming period?
It is important to clearly understand what these objectives are as they could provide a common basis for all other initiatives and activities. It also provides

guidance on how to measure progress and achievement towards the specified objective.

What are the personal objectives and expectations of the key stakeholders for the coming period?
Understanding the personal objectives and expectations of the key stakeholders will provide a better insight in their decisions and personality. It will also provide an insight into the difference between personal expectations and the formal objectives for the organization. Organizations without a formally communicated strategy may, in many cases, have its objectives dictated to it either by these personal stakeholder objectives or by a higher organization entity.

What is the key strategic proposition: operational excellence, customer intimacy, product leadership, innovator or fast adaptor?
Understanding the strategic proposition provides significant guidance for the key decisions to be made. Many people still misunderstand the difference between these strategic propositions and the resultant impact upon the business processes. It is important that sufficient validation questions are asked to ensure that the interviewees fully understand the consequences of the choice.

What are the key issues that need to be addressed?
The purpose of this question is to understand what the main initiatives need to be. The answer will provide an idea about what are the main challenges that must be overcome to achieve the specified organizational objectives. It will also provide an insight into who are the key stakeholders to be engaged in these initiatives.

A workshop with key executives and managers can provide, within a relative short time frame, an enormous amount of information to assist in answering these key questions.

Once the questions have been answered it is important to communicate the outcomes. It is critical that the rest of the organizations understand these answers so that they are able to align their tasks with the elicited strategic direction. Another outcome from the answers will be that the organization can commence validating the implicit assumptions that have been made against the strategic direction, highlighting any differences.

It is crucial that executive management understand the benefits of formulating a strategy and how it will add to the sustainability of the organization.

No formal strategy process

Situation: There is no formal strategy process. The main reasons for this are:

- no experience due to low level of management maturity,
- underestimating the importance of the strategy process,
- a previous bad experience.

Recommendation: Strategy formulation, roll-out and alignment is a process itself. Mintzberg (2005) wrote that there are various types of strategic processes. Many of the frustrations and confusion in strategy formulation relate to the fact that the stakeholders involved have different views on how the process should be conducted, without actually realizing it, hence the importance of specifying (documenting and agreeing) the process to arrive at the strategy. Just as with any other business process, it is important to fully understand the objectives, steps involved, prerequisites, people required and their skills and availability.

Unspecified impact of strategy on business operational processes

Situation: These organizations have executive management who make strategic decisions and expect that middle management will be able to determine the impact on the business processes. This is often the case with mergers and acquisitions, where many executives, for example, only address IT integration and do not address the business process impact.

Recommendation: The main challenge is to convince executive management about the need to be more process-focused:

1. Determine the desired specific element of the strategy and then find business processes that are in conflict with it. For example, an organization that wants to follow an operational excellence model with low margins might have business processes that are either based on customer intimacy with substantial time and cost required to support the individual customer requirements, or the customer service representatives have too much authority to make reimbursements to clients, eroding or nullifying the limited margins.
2. Measure the impact of this discrepancy using ball-park figures. The total impact is determined by the number of business processes, the frequency of the business processes and the impact of the discrepancy. Most managers are shocked once they see the magnitude of the inconsistencies in the organization.
3. Organize an executive workshop to determine rules and guidelines on how the processes should align with the strategy. Focus on the key areas, from Step 2, and obtain the evaluation from the participants regarding the outcomes and the benefits from the alignment. Document these benefits.
4. Ensure that the findings of the executive workshop are captured and included in a process architecture. Assign a business process steward to 'own' it and have a basic governance model in place. Ensure frequent monitoring of the specified benefits from Step 3.

Misunderstanding of the current strategy

Situation: The organization and its staff have an incorrect view of the strategy. As a consequence management makes the wrong decisions regarding business processes, we have experienced the following situations:

- Misunderstanding of the operational excellence strategy, such that:
  - *Operational excellence is not customer oriented.* In fact, in all strategic choices the business processes should be customer-focused, but

they vary in the manner by which they provide value to the customer. Operational excellence intends to provide the customers with the best value.

The business processes in operational excellence should be focused on providing the customer experience in the most efficient way. This could include the decision not to support all customer requests.

– *Operational excellence is boring and monotonous.* All strategies require empowerment of staff to ensure correct execution of the business processes. Operational excellence, just like all other strategic choices, requires full commitment and participation of staff for continuous process improvement.
– *Operational excellence is only for the back-office.* Strategic choices relate to the entire end-to-end business processes of an organization. It is impossible to have a more than one strategic choice being applied to the one end-to-end business process (e.g. one in front-office and one in the back-office).

- Misunderstanding of the variances with existing strategies:

  – *Dual brand strategy.* It is important to realize that an increasing number of organizations have multiple brands, with the different brands deploying various strategies and strategic choices. A famous example is Fiat which produces both Fiat and Lancia automobiles both with an operational excellence strategy, while Fiat also produces Ferrari's and Maserati's, which use a product leadership strategic choice. It is important that business processes align their execution with the strategy it supports. Some organizations have all the business processes for dual brands completely separated through different business units and companies, other organizations have some of the processes combined in a shared service facility. It is important to find the right balance between efficiencies and sufficient differentiation in the products and services.
  – *Strategy shift during the course of the product life cycle.* The strategic positioning of products and services can vary throughout their product life cycle. Many products, especially the innovative ones, commence with a product leadership strategic positioning at the start of their product life cycle. They may start with an initial high price to attract the 'innovators' and ramp up production to attract larger group of customers – the early adaptors, the early and late majority. Then strategic positioning may change to more operational excellence as more advanced products and substitutes have been introduced. The business processes need to support this shift in strategic positioning. This often occurs because of economies of scale and the reduction of required agility and flexibility.
  – *Customer choice.* An increasing number of brands provide customers with the choice of the value they obtain and the price they pay for it. A good example is the differentiation between first

class, business class and coach (economy) class in the airlines. Although this differentiation has existed for many years, there has been more of a shift to differentiate the business processes relating to these services. Thus, business processes supporting the same brands need to distinguish the various types of customers they support.

Immature process architecture

Situation: the organization does not yet have a process architecture and therefore is unlikely to have an overarching enterprise architecture. It is recommended to apply the Architecture Maturity Model (Van Den Berg, 2006) to demonstrate the organizational maturity in this area and then to gradually increase the level of maturity.

Step 1: Determine current level of Process Architecture Maturity

Step 2: Determine the required level of Process Architecture Maturity and specify the benefits associated with that level of maturity. It is recommended to determine a multi-year vision with a high level cost–benefit analysis.

Step 3: Develop a vision of architecture, ideally with the process architecture embedded in the overall enterprise architecture. The vision document needs to contain the following elements:

- position and contribution of architecture to the organization
- elements of the architecture
- architecture governance.

Ideally this vision document should be limited to two or three pages and should be able to be understood by all key stakeholders.

Step 4: Develop a roadmap for architecture to ensure there are regular checkpoints or milestones to determine the progress, as well as the benefits achieved from the architecture maturity.

High level of architectural thinking with limited organizational embedding

Situation: The architecture team has developed an elaborate architecture, however this is not actually being used within the organization. This is often because people find it either to be complex to understand or that it does not add value to their work. In many cases the architects are frustrated that their products and services are not being used and rather than address the gap in communication they often want to further improve their architecture.

The most critical aspect in this situation is to start the right dialogue between the architects and their key business stakeholders. The level of the stakeholders should be of sufficient seniority to ensure that the agreed business processes and rules are actually able to be enforced.

Recommendation

Step 1: Assess the current architecture, architecture governance and architecture approach.

Step 2: Determine the key stakeholders and specify their requirement for architecture in the organization.

Step 3: Organize an executive workshop agreeing on the position, role and benefits of architecture, the current level of architecture maturity, the level required and the way to achieve it.

Step 4: Select pilot projects where simple versions of the architecture will be used and show how they will contribute in specific projects. This can then be evaluated and the contribution of the architecture is determined.

Step 5: Introduce Project Start Architecture (Jeston and Nelis, 2008) for each new business process improvement initiative.

Step 6: Develop a roadmap for embedding architecture in the organization.

Chapter 8

People capability

Talent wins games, but teamwork and intelligence wins championships

(Michael Jordan)

Introduction

People are the centre piece via which a process-focused organization is created and sustained (Figure 8.1). In this chapter we will show that the building of internal capability within an organization is critical and must be supported by the creation of a Center of Business (Process) Innovation (CBI). Both will only work effectively if the engagement model between the business, CBI and

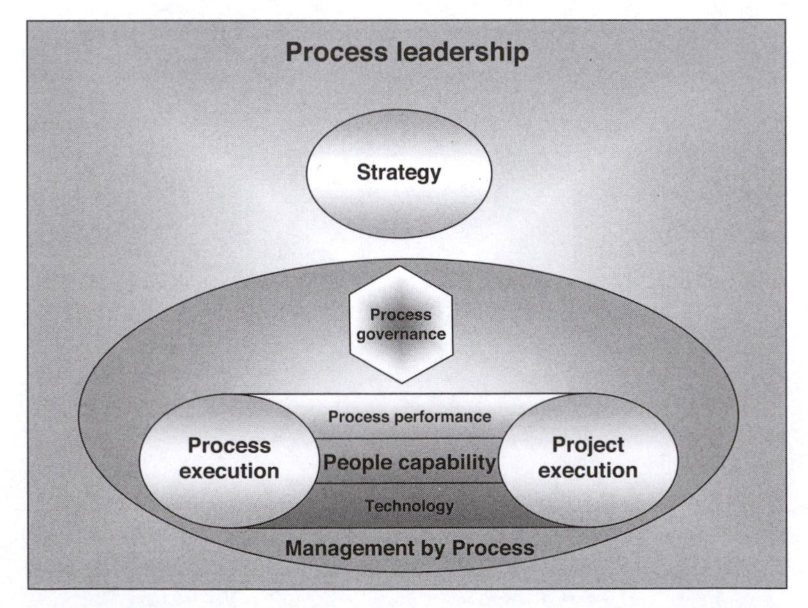

Figure 8.1 Management by Process framework: people capability.

information technology is appropriate and works. These assists in the building of the road map towards a visionary organization for the people capability perspective.

Why is people capability important?

When striving towards a process-focused high performance management organization the people are a critical component in execution. Hammer stated it well when he said (1993), 'Coming up with the ideas is the easy part, but getting things done is the tough part. The place where these reforms die is … down in the trenches' and who 'owns' the trenches? You and I and all the other people. Change imposed on the 'trench people' will not succeed without them being part of the journey.

> Forceful leadership can accomplish only so much. The shift from machine-age bureaucracy to flexible, self-managed teams requires that lots of ordinary managers and workers be psychologically prepared.
>
> (Hammer, 1994)

The most important component in any business process-focused change is the management of organizational change, the associated people (staff) impacts and providing the skills and capability for the staff to be able to execute their jobs to a high standard. Execution, whether project or process, is owned by the people in the trenches. People and their engagement is a critical factor and a holistic approach is essential. The key to engaging the people in the trenches is leadership from their line managers and the line managers must be engaged first.

It is the people who will determine the success (or otherwise) of your process-focused programme. People need to be included as an integral part of the development journey. They need to be consulted, listened to, trained and communicated with on a regular basis. If they do not understand the reasons for change or the purpose of the organizational strategy, how do you expect them to take any ownership and responsibility?

People need to understand clearly what is expected of them and how they fit into the organization. Their performance measures need to be developed in consultation and agreement with them – so make them part of the solution.

Key trends

While there is a growing awareness of the importance of, and benefits to be gained from, becoming a process-focused organization, it still has a long way to go before it is universally acknowledged and accepted. It is a rare CEO or senior executive who will focus their organization upon business processes. It is an interesting phenomenon that many consultants, vendors and indeed staff 'get' the fact that business processes will make a significant contribution to an organization (just read the case studies in this book for evidence) and yet there is reluctance from CxOs and other senior executives to embrace it. This is one of the topics we discussed in Chapter 4.

There is also a growing desire from organizations (as there should be) to build internal capability. There is an unwillingness to bring in large teams of consultants and hand over responsibility for business process improvement (BPI) and this is a good thing in our view.

However, many of these organizations do not have the internal capability and skill sets that are necessary to develop this internal requirement and yet there is an unwillingness to engage with external expert(s) to provide assistance to the organization in building the desired internal expertise.

We have also had organizations and respected colleagues say to us that the role of the chief information officer (CIO) has a limited life and with the growing understanding by the business of technology this role will change in the near future. One of the reasons for this, in our opinion, is the fact that there is a growing understanding that not all IT departments are providing a business partnership and the expected level of added value to the business. This must change and a genuine 'partnership' engagement model established.

Key elements of the people capability

We will now outline how to build internal business process capability and why it is necessary; we will cover the question, is a CBI a temporary or on-going requirement within an organization and when is the 'right' time to establish one; and how the business and these groups should ideally be working together in an engagement model that will benefit the organization by adding significant value.

Visionary people capability

The visionary state for the people capability dimension is an organization that is so aware of its business processes that they are in the forefront of people's minds at all times. However, this alone is not enough. The people within the organization must also have the knowledge and skills to be able to continuously improve the business processes, and measure and manage them in such a way that it leads to the betterment of the organization.

While people, particularly executives and senior managers, are often self-centred and concerned with their careers and power base, the visionary organizational culture will not tolerate this behavior if it in any way impacts upon the organizations performance.

> A *leaders* 'first duty is to his people. He should take care of them with no thoughts of pleasing himself, subordinating his own wishes and desires to those of the people. He should guard them as a mother guards the child'
>
> (Hawley, 1993, pp. 182–183)

Oh, to work in an organization that totally lives these values!

As any organization is only made up of people, the *people are the organization* and they need to be nurtured, trained, coached, mentored, managed and cared for.

While the organization needs to care for the people, the people need to care for the organization. One of the main ways people care for an organization is to contribute or add value to it and this will mean having their contribution measured. Performance measurement, including personal KPIs and targets, will be constructed in such a manner that they are all directly linked to, and intertwined with, the organizational strategy and continued development of the competitive advantage of the organization (as discussed in Chapter 6).

In the visionary organization there will be less need for a CBI because the expertise and skills that were resident within this group are now spread throughout all levels of the organization. Similarly, IT will understand that its existence and role is to fully support the business and will be genuinely considered a full business partner and an integral part of the organizations success. The business and IT divide will be considered a thing of the past.

Roadmap to people capability

We will now turn our discussion to how the organization may continue its journey to this visionary state and firstly consider the building of the internal capability that is considered essential. Figure 8.2 shows the steps applicable in the people capability dimension.

Building internal capability

Understandably organizations would rather, and should, build internal capability to successfully complete process improvement projects and then

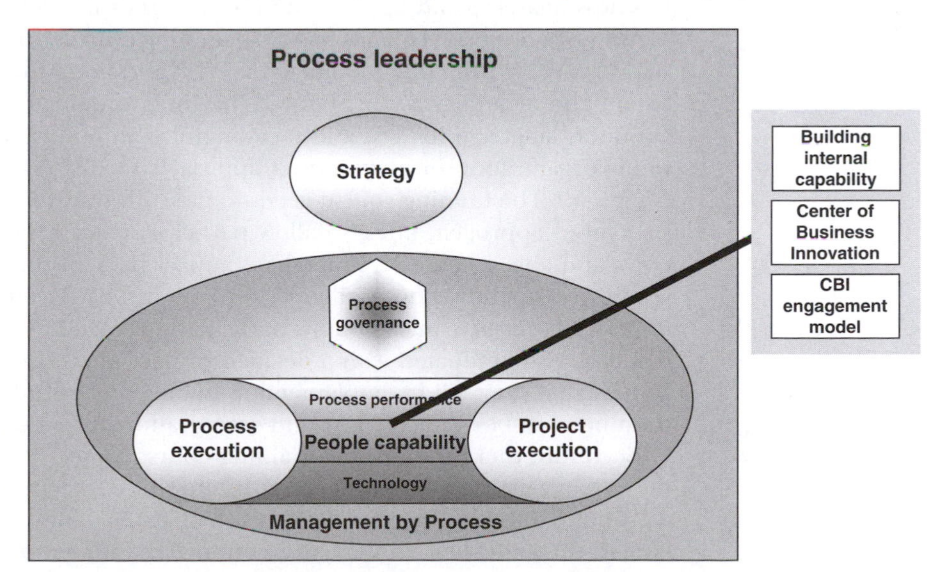

Figure 8.2 Management by Process framework: People capability steps.

sustain a focus on the on-going management of its business processes. This is the correct course of action for an organization in the long, and sometimes even the short, term. Each process improvement project must add value to the internal knowledge base and skill levels of employees.

The question is: 'how can internal knowledge and skills be built when many organizations are either:

- starting from a base of no internal knowledge or skills; or
- have an inadequately low level of internal expertise; or
- attempt to 'buy' this knowledge and skills by employing (or contracting) new staff who are supposed to have the necessary skills?'

The challenge goes beyond just the building of the internal knowledge and skills, into a complex web that includes all the aspects of each dimension outlined in this book and the need to ensure each of these aspects are available and at the right level for the organizations current business process maturity and journey.

There are a number of structured steps that an organization can take and these include:

- training courses
- provision of individual and group coaching and mentoring
- selective use of expert consultants.

We will now take each of these and discuss them in more detail.

Training courses

One generic business process training course for an organization will not be enough to build internal capability. The training will need to cover the complex and unique requirements of each organization and the various stakeholder audiences. We see these needs falling into a small number of training categories.

Firstly, there will be a need to train project managers, process analysts (or business analysts) and appropriate business staff in a structured way of completing BPI projects. The reason for suggesting a structured approach is so that staff may move from one project to the next and bring their expertise in a consistent and meaningful way. Staff will 'hit the ground running' and also have a common language for fast and meaningful communication. Further benefits to an organization of a structured approach to BPI projects are outlined in Chapter 9. The training course needs to not only transfer the knowledge of the structured approach, but also allow participants to make their learning concrete and able to have an immediate impact back in the business. When you read the case study about Citibank Germany, its 'Advanced' training course not only taught participants about its particular approach to BPI projects, but participants also graduated with a detailed project plan for their 'break-through' project that they will be implementing immediately. So the learning from the training must be both concrete and organizationally relevant in delivery.

Secondly, there should be a training course aimed at a higher level within the organization. It needs to aim at the business executives who manage BPI activities, and programme directors/managers. This should be a shorter course than the first course, and conducted with more of a business focus providing a high level overview of the BPI project structured approach and orientated at how these senior managers should be both managing the projects and the project managers, and how the business processes should be managed within the business as usual environment.

The third course would be aimed at the business process stewards/executives and how they should be executing their role and responsibilities. This is about the operations side of the business and the management of the transactional execution of business processes. Again, Chapters 5 and 6 has already provided an insight into the role and responsibilities of process stewards/executives. Process stewards are a critical role within an organization and the need to ensure that they clearly understand and execute their responsibilities is crucial. The relationship between the various process stewards is also critical to process-focused organizations.

The last course could be a short two to four hour event aimed at both senior executive and staff. The purpose of this shorter 'presentation' is to provide the audience with an understanding and appreciation of the importance and opportunity that having a process-focus presents to the organization. Once again, refer to the example in the case study of Citibank Germany see Chapter 2. The content may be orientated slightly differently for the executive audience and the staff.

Each of these courses has a different purpose and is aimed at a different audience within the organization. Unless each audience is addressed, the organization will not progress in its process-focused journey and reap the available substantial benefits.

Coaching and mentoring

Unfortunately most organizations believe that all they need to do is to train staff and they will then have the skills and knowledge that is required. In our experience this is not always the case, in fact, it is rare. Training is the appropriate starting point to establish foundation knowledge, but unless supported by on-going coaching, the training investment will rapidly diminish with time. The knowledge gained in a training course can only be generic in nature, no matter how specific to the organization it is. Coaching and mentoring allows staff to be assisted in applying this generic training in each unique situation.

Case study: Large university BPM training and coaching experience

This organization had two failed project implementations and decided that one of several primary reasons for the failures was that it did not truly understand their existing business processes. It commenced a BPM project to model and understand the current business processes. The university, and particularly the project manager, recognized that they did not have any BPM skills and knowledge within the project team. To reduce the project risk and build internal expertise the project manager engaged an experienced BPM expert to coach himself, the project team members, provide project quality assure and assist with the direction of the project.

(Continued)

Case study: Large university BPM training and coaching experience (*Continued*)

BPM training was provided to educate the project team in a BPM framework, or structured approach. The team was then supported with coaching. The coaching took the form of:

- one-on-one coaching of team members;
- for specific workshop facilitation training, the coach acted as the facilitator in some of the initial workshops to model the skills for the team members; the team members then took over the facilitation role with the coach being able to provide constructive feedback, until the team members were confident they had acquired necessary skills and knowledge to complete these tasks themselves;
- assistance was also provided in the steps to be completed and the review of the activities and resulting documentation.

This process culminated in the successful completion of several business cases for the improvement of the universities business processes and their subsequent implementation. Some of the improvements showed productivity gains of in excess of 47% and all were in excess of 30%.

Message: The risk on the project needed to be minimized, which is one of the key functions of any project manager. This was especially true of this project in the light of the failed project history. Expert BPM project coaching was considered a key activity in this risk mitigation strategy and was exceptionally effective.

It is generally understood and accepted that athletes, even the world's best athletes, need coaches. It is also understood that these coaches can, and generally do, make a significant difference to an athlete's performance. Roger Federer, the world's best tennis player, pursued Tony Roach for months encouraging him to be his coach, and the rest is history (although they parted company during 2007).

Coaches have the advantage of looking from the outside in and looking for finer distinctions in a person's performance. They can teach a person about aspects of performance that the athlete does not know, and will also assist with mental attitude and approach.

So why is it any different in business? After all, as Andrew Spanyi (2003) states, 'BPM is a team sport and sport is about winning' and in sport, there is no discussion, we have coaches. Why is it questioned in business?

Training alone is not enough without the benefit of a coach to support the training exercise. Do organizations need a business process or BPM coach? Do specific individuals within those organizations need a coach, and, if so, why?

We have observed many organizations and how they have approached BPI projects from an implementation perspective and in the creation of process governance and management structures. We see the approaches that have been successful (what worked well) and the mistakes made (and what could have worked better).

If we were to generalize, organizations take one of three approaches to business improvement projects or programmes:

1. Organizations 'have it done to them' – it brings in an experienced BPI consultancy or vendor to take responsibility for the delivery of the entire project(s);

2 Organizations 'do it to themselves' – it does not need any outside help (or need only a minimal amount) as it can determine how to do BPI for themselves; or, finally,
3 Organizations 'do it to themselves, with expert coaching and guidance' – selective use of expert assistance to help determine the best way forward for the particular organization and provide coaching to the process management, project managers and project team staff.

One could argue simplistically that each of these approaches is related to the maturity of the organization and its understanding of the relationship of business processes to success and high performance management. The first approach represents the least mature organization (assuming it does not have any previous experience with BPI) through to the third approach where it is understood that the organization does not have sufficient internal expertise or capability, and requires assistance to move along the business process maturity journey.

We will now take each of these and discuss why an organization may adopt a particular approach, and the respective advantages and disadvantages of each.

'Have it done to them'

Why would an organization want to have another organization come in and take all, or most, of the roles in a BPI project and be responsible for the outcomes? We have experienced the following reasons, and are sure there are many more:

- The organization has no idea what to do or how to approach a BPI project. It does, however, understand it has some serious operational issues, such as not meeting expected service levels to customers, unacceptably large backlogs in processing, high levels of overtime being paid or poor levels of customer service.
- This approach is perceived as a low-risk option by the business: 'Let's get the experts in to do it'.
- The organization has a culture or expertise in outsourcing, and this is seen as 'just another' outsourcing deal.

There certainly are some advantages associated with this approach. A couple of them include:

- Faster implementation – if the BPI implementation experts are running the project, the likelihood is that the project will be implemented faster than otherwise would be the case. Mind you, a fast implementation has its own set of risks associated it.
- If the external parties are directed correctly by the engaging parties within the organization, they will be very focused on the delivery of the project benefits.

One of the major difficulties, or disadvantage, with this approach is that the organization could behave as if it had totally 'outsourced' the project and it does not need to take any responsibility for the delivery of the project, or does not need to take part in the project at all.

This is a recipe for disaster as the business must be involved in BPI project, as BPI is all about the operational 'management of your *business*'. In fact, the business must be intimately involved in all aspects of the project. There are decisions to be made by the business, and essential business knowledge to bring to the project; and when the project is complete, the business will be responsible for running the new or redesigned processes, not to mention their on-going improvement.

In fact, if the business has not been significantly involved in this type of implementation, the risk of failure is extremely high. The external experts may have knowledge of BPI techniques and the implementation of these projects, but they are *not the experts in your business.* How will the new or redesigned business processes align with the organization's strategy? How do you know that the staff will accept the new business processes if they are not engaged in their creation? Who will align the new business processes with the process and staff performance targets?

This approach will also mean that it is unlikely that there will have been much transfer of intellectual property (knowledge and skills) to the organization staff and therefore the building of internal capability. As the business changes and the business processes need enhancement, the external 'experts' would need to be re-engaged to repeat the project.

In conclusion, this approach is not recommended as the involvement of the business is an essential ingredient in the success of any BPI project. The experts may solve the wrong problem and not address all the issues associated with the original operational challenges or the reason for the project in the first place.

'Do it to themselves'

Why would an organization adopt this approach? The number of reasons is probably as long as the day is long and will often include:

- We are unique – there is no other organization quite like us and, therefore, bringing in an external organization or consultant to assist would not help because 'they just would not understand our particular situation, environment or culture'.
- We have intelligent people working in the organization; they are smart enough to figure this out for themselves. They can learn on the job, as they go.
- We just do not like consultants – we have had some bad experiences in the past; they cost a lot of money and are simply not worth the money.
- If the organization was being honest with itself it might have a 'not invented here' perspective. That is, the organization does not respect or value external knowledge and only values the knowledge it has created or gained itself. Sometimes this is related to the egos of members of the staff or management.
- We can save a lot of money by doing this ourselves.

Some of the advantages of this approach will include:

- A sense of satisfaction that, once achieved, the organization has done it by itself.
- The lessons learned will be hard won and therefore memorable. This will mean that the lessons will be remembered by the individuals and, hopefully, by the organization as a whole.

- The intellectual property gained will be retained within the organization, for the reasons mentioned above.
- The solution will be specifically tailored to meet the specific organizational requirements.

The disadvantages could include:

- 'Doing the "hard yards" yourself' within an organization will mean that mistakes and dead ends will be encountered. While these may result in memorable learning, they are also costly to an organization in terms of time and effort (and therefore money). Things will simply take longer, and often not be performed as well as it could have been.
- The inexperience of the participants in these projects will mean that the risk of project/programme failure will increase. One of the key aspects to project management is the management and minimization of risk. 'Doing it to yourself' will not help minimize this risk and potentially significantly increase it.
- The increased risk will potentially lead to project failures. The challenge with this is that the failure may be perceived as a failure of BPI projects and a process-focus itself to deliver the benefits promised, thus leading to the cancelation of an entire BPI programme.
- The above points will most likely lead to a delay in the realization of the business benefits associated with BPI. This will have both monetary and business agility impacts that may be unacceptable to the business.

In conclusion, we believe this approach is false economy unless the organization is experienced in BPI with multiple projects completed successfully. The inevitable delays in time and benefits realization, mistakes and higher risk factors will be expensive to the organization and place the project at a higher risk status. Some organizations try to overcome these issues by employing one or more contractors who claim to be BPI experts or have BPM experience. It is not only 'some BPM knowledge' that is required. There is a need for a structured approach, together with deep and experienced implementation knowledge and process coaching skills.

'Do it to themselves, with expert coaching and guidance'

The organizations that adopt this controlled approach are usually more mature in their culture, leadership, and understanding of business and process-focused knowledge.

These organizations understand that:

- Selective use of business process experts to assist the organization can significantly help in the successful implementation of BPI projects, the creation of a process-focused and high performance management organization. The organization understands that it 'does not know what it does not know,' and the experts will help assist with this.
- A coach can be a role model, especially for project managers and project team members. For example, they could facilitate some of

the initial workshops to demonstrate the skills required, and then progressively have the project team members take over.

- Risk minimization strategies are critical to enhance the prospect of success and avoiding failure. Expert coaching will significantly assist in this risk reduction strategy.
- The organization cares for its staff and recognize when and where staff need assistance and support to help make both the staff and the organization successful.
- It respects its staff enough to support them in areas they do not have a high skill level in or do not understand well. They do not see this as a weakness in staff, just that staff require some training or coaching to overcome the lack of experience and to provide increased skills and knowledge.

The advantages of this approach include:

- The ability to implement the project/programme faster than other approaches.
- Risk is minimized in comparison to other approaches.
- An understanding that business benefits will be realized faster. These benefits are not the only reasons for the project (examples, as shown in the business case, include increased customer service, reduced backlogs and increased turn-around times); business benefits will also increase business agility, that is, the organizations ability to react to competitive pressures and take advantage of market opportunities.
- The transfer of intellectual property to organization staff will be much greater. In fact, projects should be conducted so that the coach(es) support the organization's staff. The staff should take the lead in projects, while being supported by the coach.
- The cost is actually less than without the specialist coaching – when all these considerations have been taken into account.

Disadvantages include:

- The initial cost of the external coaching support may appear, or actually be, more expensive in the short term and will obviously be reduced significantly over time as organization staff gain knowledge, skills and experience.
- There may be a fear of 'outsiders' coming in to the organization and embarrassing staff, but this is counter to the role of a coach. The coach should clearly be in the background. The coaches role is to make the organization staff successful (the 'hero').

In conclusion, the use of this approach by organizations shows that they have an understanding of the difference between their *capacity* to implement BPI and their *capability* of doing so successfully. They understand that they may have the resource numbers to complete the project or programme, but do not have sufficient skills, knowledge or experience to do so in the desired timeframes, ensuring the realization of the needed business benefits.

So let us take a few moments and determine just how an expert business process coach could assist an organization in the implementation of a BPI project or programme of work and review some of the possible coaching activities. The coach could also assist business managers, process executives and process stewards in their roles within the business.

Typical services that could be provided by a coach will obviously depend on a particular organization's needs and could include:

- Assisting business managers in an understanding of how to manage their business from a process perspective.
- Assistance in creating an approach to a project or a programme of work that will yield predictable outcomes and reduced risk.
- Stakeholder identification and management needs to be approached in a structured and analytical manner.
- Staff coaching for a specific project or programme.

We will take each of these and discuss how an organization may engage with a BPM coach.

Assisting business managers become high performance managers

This basically covers all the aspects described within this book. The coach will be able to bring a different and wider range of experience in the execution of the various suggested roadmap activities. The implementation of each dimension and activity will need to be considered in the context of the organization and executed accordingly, however, this does not excuse the manager from the need to implement it. Managers also can become overwhelmed with the day-to-day activities and problems with their role or business unit and a coach will enable the manager to maintain a 'big picture' view and put things in perspective.

Creating a structured approach for predicable outcomes through BPI

The coach can assist in the creation of the structured approach (framework/methodology) for the management of business processes and BPI projects. Does the organization require an approach that is 'strategy-driven' (part of the implementation of the organizational strategic plan or vision); 'business-issue driven' (needed as part of addressing an existing operational problem) or 'process-driven' (service level motivated)? The business drivers will impact the approach to the initial stages of the business improvement projects. Are the drivers obvious or hidden? Is the project scope clear? Are all the stakeholders in agreement with it? Has it been written down and agreed upon? The coach will assist in 'fleshing these out' and gaining agreement.

How does the organization select which business process(es) to address first and in which order to maximize the investment? How important are business process metrics and how much detail is required to proceed? Once it is decided that a project is to be established, the coach will assist with optimizing the structure of the project team, establishing the roles and responsibilities and so forth.

Are the business process performance targets appropriate? Are they too detailed or not detailed enough? Is the level of process governance sufficient and are the roles and responsibilities clear, understood and accepted?

Traditionally, projects delivered under project management methodologies have targeted the technology-based or low-resistance business changes. This enables a certainty of delivery that project sponsors like to have. BPI changes are different in one very important aspect: they almost always require a large element of people and/or cultural change management.

Why is this? Consider any manager in control of inefficient business processes in a large organization. Nine times out of ten, the manager is well aware of the inefficient operation – as indeed are his or her superior managers and most of the processing staff. Once a business process expert is brought in to improve the situation and recognizes that stakeholders understand that the business processes are inefficient, the obvious assumption to make is that the implementation job will be easy. This, of course, is not the case. The difference with business improvement projects is the degree of impact upon the people and culture. The underlying causes, interests and agendas that require the business operations to change are often not recognized, understood or addressed. We are now deep in territory where a traditional project manager does not want to be. This world is uncertain, high risk and not easily controlled. An experienced business process project manager understands this environment and is skilled enough to succeed in this type of environment. The expert process coach can assist the project manager with these challenges of the project.

Stakeholder identification and management

Most people, managers and project managers handle their stakeholders by intuition, and, while stakeholder management is an integral aspect of any business and project, the in-depth knowledge and experience required in a process-focused environment (and process improvement project) is much greater.

Stakeholder management is all about relationship management and is a structured process-based approach for handling the necessary relationships involved. Owing to the complexity of BPI projects, this stakeholder management needs to be a more formal process than in traditional projects.

How do you create this more formal stakeholder management structured process? There are two types of stakeholder management usually required for successful BPI projects. The first is called 'managing stakeholders for successful delivery.' The second is 'interest-based' stakeholder management, and this is based on co-operative problem-solving techniques. This is where relationships are made and maintained that progress towards permanent change in individual and group behavior – more conducive to cultural change. Both techniques will need to be used for the significant organizational change that is necessary for BPM projects (Jeston and Nelis, 2008, Chapter 24).

Neutral persons (coaches) can add significant value as they can maintain objectivity and do not become involved in the politics of the organization. Often, a neutral or external person will also be listened to more than an internal person.

Staff coaching for specific project/programme requirements

BPI staff and managers can receive various process specific training courses, and, even if the training is especially designed for the organization, there will always be a need for follow-up – often one-on-one coaching. No training course can deliver enough information to enable staff intimately involved in BPI projects to be able to conduct the projects well without further support. They will need someone to bounce ideas off, someone to review their progress in a non-threatening way and someone to tell them if they are straying from the 'best road forward'. The coach can also maintain a consistency with all stakeholders and project staff.

Staff being coached are often more comfortable with an external person who will not judge them and will only provide help when they need it. Internal staff may compare staff members with each other or may report performance to superiors, when it is knowledge and guidance that is needed, not performance evaluation and reporting.

In conclusion, BPI implementations and projects are difficult, and the level of experience required to deliver successful projects must not be underestimated. 'Runs on the board' matter and having completed previous BPI projects makes a difference in being able to replicate success and guide others to the same outcomes.

Managers are continually required to achieve ever increasing levels of performances and profits. They are forced to look for new and improved ways of conducting their businesses, which often means looking at the latest management trends and fads. Some managers will purchase a BPM tool and think it will solve all their problems. They hand it over to IT departments to implement and are surprised when it is not successful or not as successful as it should be.

Having experienced BPI/BPM experts who understand this, and are able to coach both management and staff in all the aspects of complex projects and programmes and the development of a high performance management environment, can save money and improve the speed of implementations and benefits realization.

Roger Federer can have all the natural talent in the world but without practice and the assistance of a neutral person (coach) with different experiences who is willing to provide honest and constructive feedback, Mr. Federer's progress will be more difficult. It is only by honest constructive comments and suggestions that any of us can improve.

Getting to number one in the world is not easy, but staying there is even harder. You need someone to keep you sharp and motivated if you are to continue to aspire to a higher level of achievement.

Not everyone can get to be the world's number one. It is about being the number one in your selected area or the best you can be. As Andrew Spanyi has stated, 'BPM is a highly competitive team sport, and in the playoffs for industry championships, it's win or be eliminated' (Spanyi, 2003, p. 168).

Where is 'your' ambition level? A coach will keep you focused, keep you aiming in the right direction and a little higher all the time.

Continual development of intellectual property

The old saying: 'you do not know what you do not know', is not only relevant to the need for training and coaching, but also the continual development of knowledge. Many organizations become insular, with its knowledge only expanding by its' own internal experiences and the experience of the people inside. Some organizations try to overcome this by periodically 'bringing in new blood', that is, new people from outside the organization who will bring in new experiences and knowledge. While in the right circumstances this can be the correct thing to do, it is not the solution to this issue.

In the world of BPI and process-focused management, the knowledge base is growing incredibly fast and there will always be a need for the continuous development of new 'practical intellectual property (knowledge)' within the organization. This will include the introduction of not just new knowledge, but also methods, tools and techniques.

How can an organization go about ensuring that there is a continuous stream of intellectual property available to its people? We have outlined several of the options below. The organization could:

1. Send its people on various external training courses. These would not only be limited to process improvement courses, but could include university courses for senior executives (MBA summer courses); and various other well credentialed courses.
2. Provide well stocked libraries from which staff can borrow books.
3. Continuously bring in external speakers and educators to challenge and provide staff with ideas.
4. Engaging external expertise occasionally on projects or investigations to assist staff and bring in specific expertise.
5. Identify various organizations to partner with. These organizations should be 'deep' specialists in an area, for example, process improvement and process management, and have a reputation for the continuous development and sharing of its intellectual property.

There is also much an organization can achieve internally. An organization must establish formal mechanisms for the gathering, analysing and sharing of lessons learned. Some of the ways this can be achieved is by:

1. The establishment of formal Post Implementation Reviews (PIR) of projects. Many organizations complete these, but few learn from them, by simple documentation and sharing.
2. After every new venture in the organization, whether a new product/service launch, new product/service implementation, conduct a review. We only ask two questions: 'what worked well'; and 'what could have worked better'. You may note the particular wording of these questions, the latter is stated in a positive way, not to imply criticism and allow staff to feel comfortable in contributing.
3. Allowing staff to be comfortable to try new things *and fail*. This is a significant challenge for most organizations, and people in life generally, as we have all been taught that 'failure' is a bad thing. Failure is only a bad thing if we continue to make the same mistake over and over again. Michael Jordan, one of the greatest basketball players of all time, is purported to have stated that: 'you make zero percent of the shots you do not make.' If you do not try, you cannot succeed. Will you 'make every shot'? Not even Michael Jordan made every shot. Developing this mindset and allowing it to percolate the organizational culture will be a significant challenge for most organizations. Do it – the rewards are amazing! Once you have discovered things that work, and could have worked better, share the knowledge and learn from it.

These are some of the many options available, and like most things in life, there is no one answer. The answer will lie in a combination of several solutions and they will differ for each organization.

The implementation of these takes courage from management and unfortunately few will dare to try. The one thing that can be stated for certain is that it cannot and will not be achieved within the organization alone.

Center of Business Innovation

In this section on the CBI we will describe: what is a CBI, why do you need one and the benefits, what it should look like, when you should dismantle it and how it should engage with the business. Later, in Chapter 9, we will make a comparison with an organization's project/programme/portfolio management (PMO) office.

Firstly, let us state that while we have labeled this as a CBI, which can have wide ranging implications within an organization, we will only describe it from a business process perspective.

The typical CBI cycle that an organization will progress through is:

1. Having no CBI within the organization and not realizing that they need one.
2. Understanding that a CBI will benefit the organization, selling it internally, gaining approval to establish it, and determining the appropriate structure, roles and responsibilities and where it fits within the organization.
3. Establishing the CBI and running it successfully.
4. Finally, scaling down it because the skills and capabilities that have been created within the CBI have been integrated into each of the business units and having a process-focus is 'just what we do around here' and therefore, the process skills now reside within the business, where they should be.

What is a CBI?

Increasingly large organizations with multiple lines of business are under pressure to reduce operating costs, create an increase in both business and processing capacity, increase service levels and enhance its governance across the organization. However, most business units will have its own groups trying to complete these activities, resulting in duplicate effort and increased organizational costs.

A CBI is a means by which an organization can centralize and institutionalize its BPI and management expertise and experience, thus reducing this duplication. It is not however, only about reducing duplication, it is about bringing together people with similar and different skills and experiences to solve complex business problems.

It also aims to facilitate co-operation between the business and IT, giving the business greater responsibility for the delivery of automated and non-automated business process solutions. The CBI is the central point for the pooling of expertise, experience and resources to assist a wide range of business units to develop, implement and/or manage process self-improvement projects and the embedding and management of business processes within the wider organization to create sustainability.

Why do we need a CBI and what are the benefits?

If a CBI 'team' is justified, then it should be established as the organizations experts in BPI and management. They should be the part of the organization to gather, analyse, document and share the lessons learned, as discussed earlier.

It will be able to assist various parts of the organization in the establishment of a 'management by process' structure and provide advice on the creation of a high performance management environment, for example, the establishment of the business key process performance – measures and targets. The team will also be able to assist individual projects with advice, coaching and perhaps resources, depending upon how the group is structured.

The challenge is to build a business case for the establishment of a CBI group in the first place and this will predominantly revolve around how it is structured, its purpose, benefits to be gained and cost. That is, what will be the contribution or value-add the CBI will make to the organization.

It is easy to write words in a business case outlining the benefits and contribution the CBI can make, but the difficulty comes in the quantification of these benefits. The CBI group is not just about completing more effective projects at reduced risk; it is also about creating a process-focused organization culture that will make a significant difference to an organization, in terms of improved customer services and satisfaction levels, enhanced staff satisfaction and creating capacity within the organization.

Research completed by BPTrends, 2007 (Survey titled *A Survey of Business Process Initiatives*, authored by Nathaniel Palmer and published on BPTrends.com in January 2007) and shown in Figure 8.3, show how organizations with a higher commitment to the process-focused journey tend to be more successful in the establishment of a CBI (they refer to it as a BPM Center of Excellence), compared to a Business Process Team and no teams at all (just ad hoc unstructured projects).

The typical benefits that can be provided by a CBI group could include:

- Leveraging any existing BPMS technology within the organization, such as, document management, workflow or business rules engines and so forth.
- Individual projects are usually part of a larger programme of work, the CBI could ensure that all projects/programmes are managed under a single governance structure, assist with project initiation and ensure that the lessons learned from previous projects are incorporated into the project. The benefits of this is that there will be lower overall overhead and governance costs (most of which are hidden) and there will be better quality assurance of projects, resulting in more successful projects.

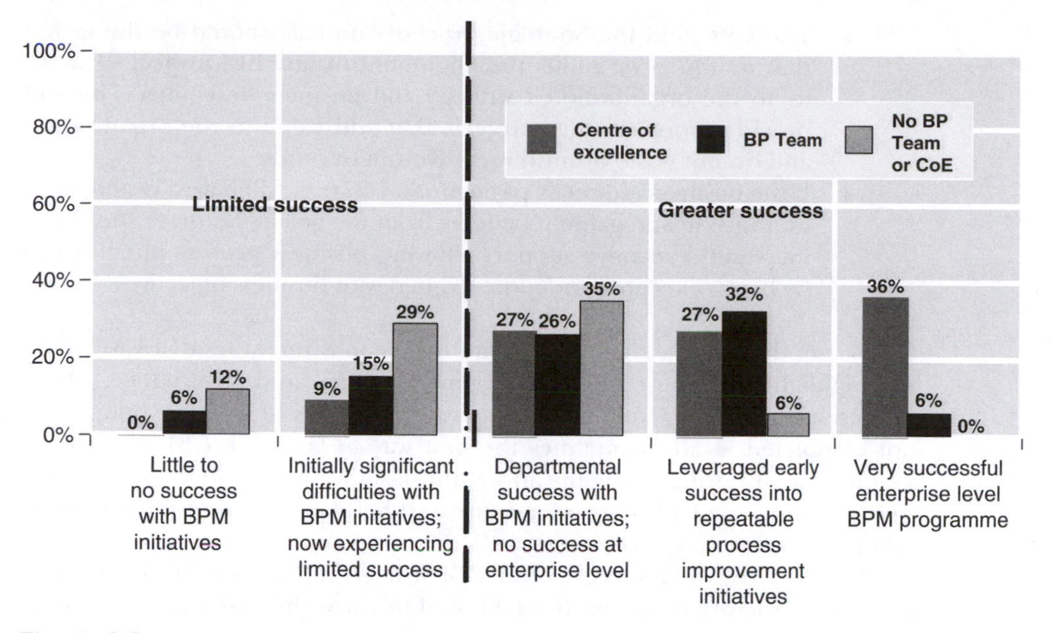

Figure 8.3
Achieving BPM success © BPTrends, 2007 reprinted with permission of BPTrends Survey titled *A Survey of Business Process Initiatives*, authored by Nathaniel Palmer and published on BPTrends.com January 2007.

- The group should ensure that if the organization requires a process modeling toolset, that it is only using one and not several toolsets across the organization and that there is a consistent and single modeling methodology being used. The benefits of this include: savings on software license fees, maintenance fees, training and management costs.
- A combined group of qualified and capable process experts will enable the optimization of their use within the organization. Experience has shown that a CBI can create a process-community within an organization where people assist each other and share experiences in process-related work. The benefit is better support for the business and projects; and lower costs of external process expert resources.
- A centralized CBI will provide the ability to work on a process-focused and continuous improvement culture resulting in more effective process initiatives within the organization.
- The CBI could provide a single programme to monitor and determine the required level of process maturity required by the organization and consistently and continuously ensure that all activities contribute towards the advancement of the desired process maturity level and more successful process initiatives. The benefit of this is that any investment in process improvement and management in the organization are sustainable and contribute towards the achievement of the required business outcomes.

- The CBI (and the Strategic Process Council) should be the group that monitors or audits the alignment of any BPI project or activity to the organizational strategy and business outcomes. This will provide more effective projects that add value to the organization and do not waste or misdirect investment funds.
- If the business is directly paying for a CBI they will have a vested interest in using and gaining benefits from the group. So there should be increased executive support allowing business process initiatives to be better executed and more aligned with business objectives.

Of course, the biggest challenge is that the establishment of a CBI will actually highlight the true cost of process management and governance, where as these activities may well have been either hidden within other budgets or not completed at all. Sometimes the best way to launch a CBI is to find a senior executive with credibility and enthusiasm to help fund the initial CBI. This will allow the CBI to gain exposure and build credibility and then bring other parts of the organization in on it.

The need for, and success of, a CBI will be influenced by the process maturity of the organization. If established to early, the CBI may have a fragile existence until the organizational business process maturity catches up.

Depending upon the organization, we have seen that a CBI may be an inhibitor to business process progress within the organization if business staff adopts a view that promotes the attitude of 'it is not my job' it is the CBI's job. However, it can often be the right strategy to allow the organizational culture, people and business process maturity to develop until the organization 'demands' that one be established.

However, once established, to ensure the longevity of the CBI group, it could develop a pragmatic charge-back system to the business so that ultimately it will not need to be funded from budgetary allocation from each business unit. This will avoid the annual challenge of going to individual business units to obtain funding. It also overcomes the issue of what to do if one or more of the business units does not wish to fund the group and the funding is left to the remaining business units.

What should the structure and roles and responsibilities be?

When establishing the initial CBI, do not be too ambitious. It is better to start small and grow with confidence and credibility. The CBI will bring together the 'right' blend of business, process and IT expertise from across the organization to ensure alignment.

One of the foundation decisions that must be made with regard to the structure of the group is the services it is to offer the business. Simplistically, there are two possible scenarios:

1. it will be a central group of business, process and IT internal experts to consult and coach the business with *process-focused activities* within the organization; and/or
2. a *resource pool* of qualified, skilled and capable experts to be placed in projects and/or within the business to complete specific tasks, before returning to the resource pool for reallocation.

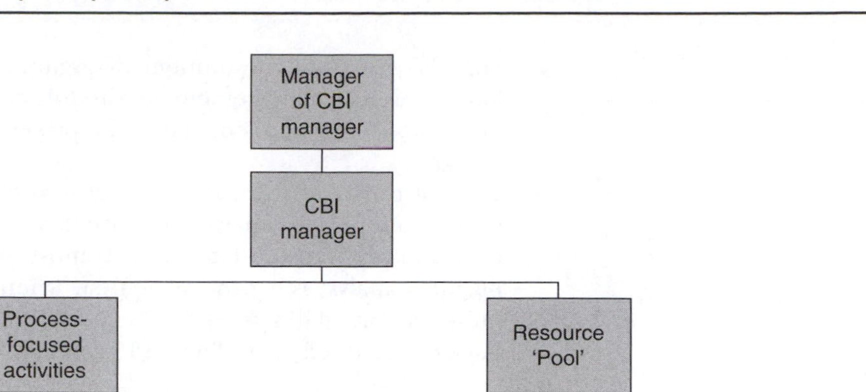

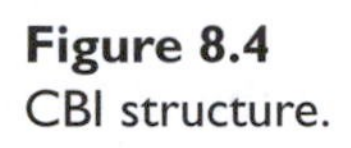

Figure 8.4
CBI structure.

Figure 8.4 depicts these two scenarios together with the typical roles in each.

The challenges that are usually faced by a CBI group include:

- The need to obtain tangible business results – to demonstrate the groups added value to the organization.
- Improve the engagement model with the business – to better serve the organization.
- Improve the project management methodology, capability and track record – to have a repeatable, successful and sustainable project methodology within the organization. This will be discussed in detail in Chapter 9.
- Improve the BPI project framework, capability and track record – by training, coaching and further staff development.
- Increase project/programme delivery capability.
- Acceptance from the business that the CBI will be the internal consultancy of choice for the delivery of process improvement, process management and change programmes.

The usual issues to be overcome can include:

- A lack of a structured and consistently applied approach to process analysis and improvement.
- The current approach to projects will be an IT solution approach.
- A lack of a uniform approach to the development or implementation of software or projects in general.

- Surprisingly to us, the number of organizations that either do not have a project management methodology, or one that is optional or inconsistently applied, across the projects within an organization is high.
- Lack of clarity of the role of process analysts across the organization. Often process analysts are either not involved within a project or are brought in too late. In fact, most organizations think that a business analyst is a process analyst, whereas, they are similar but quite separate skill sets.
- Lack of a defined or understood process architecture.

The responsibilities of some of the roles shown in Figure 8.4 are outlined in detail in Appendix B.

To the reader the CBI team must sound like a sizable group and it is the CBI manager's job to ensure this is not the case. Its actual size will obviously vary depending upon the size and maturity of the organization, however, it should not comprise any more than 3–4 or up to 10 team members, for the *process-focused activities* group. The *resources pool* group size will obviously depend on the needs of the organization and the number of projects in progress at any one time.

The CBI team exists to facilitate and support the activities of the various BPI projects, programmes and process management activities within the organization. This is why the CBI group should be centrally located within the organization, preferably at a corporate level. It must *not* report to IT, as it is a business process related activity.

Business and CBI group engagement model

We will now move on to discussing how the CBI group will need to engage with the business. It is essential that both the CBI group and the business management have an effective and working relationship. We will refer to this as the 'CBI engagement model'.

The objective of a CBI engagement model is to ensure that significant and sustainable results are achieved by ensuring that project initiatives and results are enforced as part of *management as usual*. The CBI engagement model needs to also cover the management of processes within the organization if sustainability of process improvement and process management is to be attained. It is part of the responsibility of the CPO (or process executives) to create the appropriate CBI engagement model and have it approved by the Strategic Process Council.

The main principles of the CBI engagement model are

1. *Accountability* – by ensuring that process stewards and business managers are accountable for the performance results (both quantitative as well as qualitative) of the business processes.
2. *Measure outcomes* – of both the quantitative as well as qualitative aspects of the business processes and create comparisons between appropriate business units across the organization.

3 *Enforcement* – the executive leadership must enforce the accountability of the process executives, process stewards and business managers. This enforcement task is part of the process governance and must not be placed upon the manager of the CBI unit as he/she are usually not senior enough, in most organizations, to force a difference in attitude and discipline, nor is it appropriate in any case.
4 *Facilitation* – is the main task of the CBI group. If the business executive enforces management of processes and their improvement, then the CBI group can support the business managers.

Facilitate rather than enforce

Many CBI groups fall in the trap of wanting to enforce improvements and methods onto the business. In most organizations the business is keen to deliver short-term objectives, encouraged by performance targets that focus them on the short term. Any activity initiated by the CBI group that takes them away from this short-term focus is considered a nuisance by the business. So when the business needs to make a choice between risking the achievement of their short-term performance targets and initiating longer term activities for the betterment of the organization, it will be the business short-term targets that win. If the CBI group continues to push too hard, then they will be marginalized and left to only work with a few managers who understand the importance of longer term focused business improvement activities.

It is up to executive process leaders to create an environment and culture of continuous improvement and educate managers and staff in the importance of improving and managing business processes. Executive process leaders need to ensure that KPIs reflect this business process-focus, or indeed, if necessary enforce it.

There should be comparative information available which reflect how well the various managers are performing with their key business processes. This should be discussed in management meetings chaired by the executive. The CBI group can assist and facilitate the business managers with suggestions and ideas on how to improve their business process performance. This way the CBI group focuses on facilitation and leaves the enforcement where it belongs: with the executive process leaders.

In Figure 8.5 we can see that the left side of the figure refers to the likely current situation in many organizations and the right side refers to the desirable future situation. Note: that the right side is not the visionary state, it is a step on the journey.

The current situation shows that there are many isolated BPI activities being conducted within the business. These are being coordinated and completed by isolated individual teams, which potentially limits the overall impact upon the business.

The desirable future step on the way to the visionary state has all activities coordinated via the CBI group who then facilitates these activities with the business. This provides the opportunity for a more uniform approach that should lead to a more significant and sustainable improvement within the business.

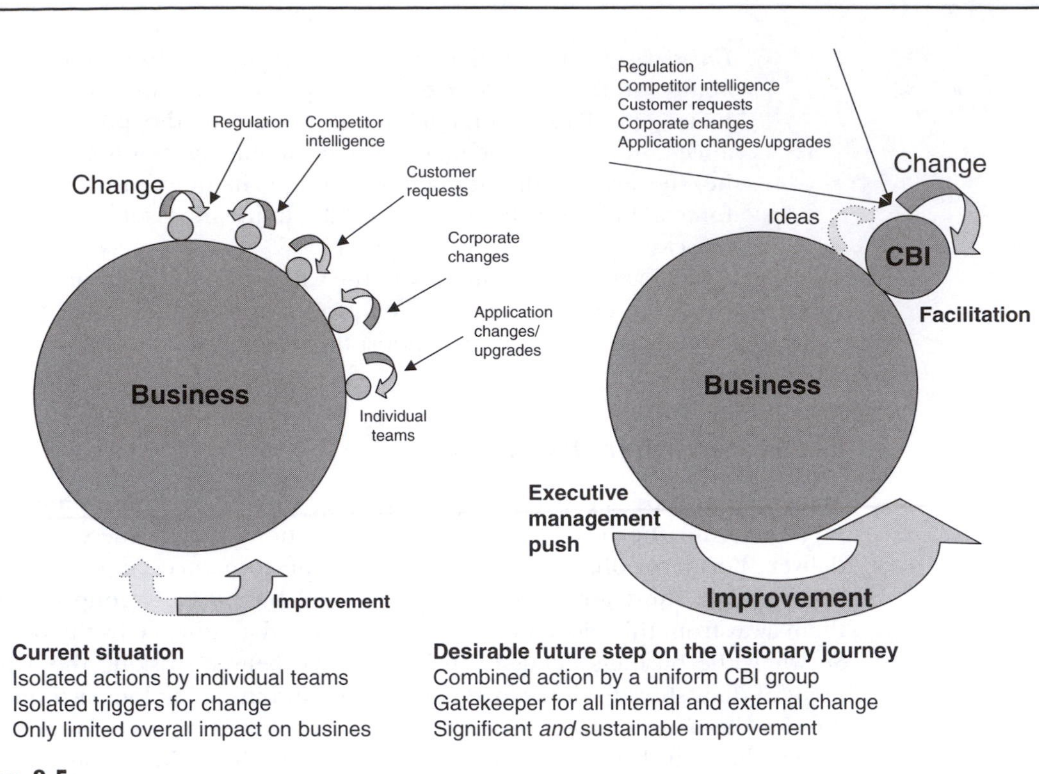

Figure 8.5
Comparative CBI/business engagement models.

Case study: Visibility drives performance

A financial organization was struggling with reducing quality, customer satisfaction and difficulty in meeting their specified targets. Analysis showed that all financial data was available at an individual staff level, however, data referring to customer satisfaction, errors, compliance, unnecessary referrals and hand-offs were only measured at the organization level and were scattered in various types of reports and memos across the organization. It became obvious that most employees only received recognition and had an ability to achieve their bonus by focusing purely on sales-related activities. This increased the error rate, increased unnecessary hand-offs and referrals (as staff wanted to focus their time purely on sales generating activities) and this impacted customer satisfaction.

We developed a scorecard which included both quantitative and qualitative targets and it was summarized into a Red-Amber-Green Scorecard. Suddenly the organization, for the first time, had a clear view on the areas and business units that required attention and improvement. The data showed clearly that several business units scored poorly across all qualitative aspects. With this scorecard now being discussed at the executive management meetings, the exposure ensured that the concerned business managers were keen to understand which of their business units were responsible for their poor overall rating. Improvements occurred almost immediately as the organization now knew which areas and business units to target for their improvement activities. Previously the organization had initiated vague and broad improvement initiatives that no one felt compelled to act upon.

Message: Ensure that measures and management information is available at the appropriate level and to the appropriate people within the organization. Also all performance measures must be visible within the organization.

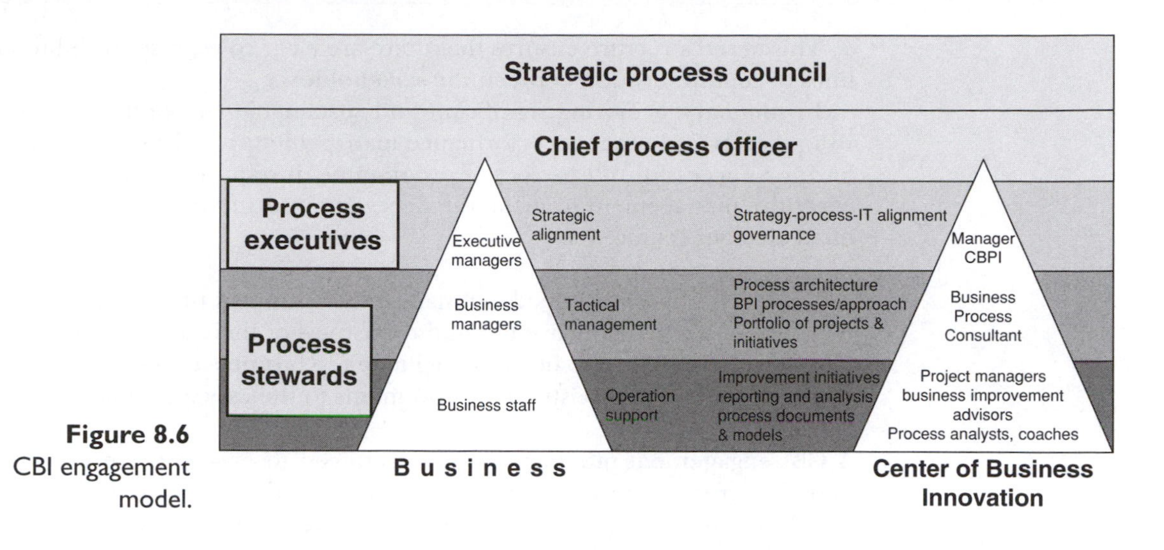

Figure 8.6 CBI engagement model.

BPM engagement with the business: Who does what?

Many organizations struggle with a structured engagement model between the business, IT and processes. They struggle with 'who should be doing what'. Symptoms of this struggle include:

- Executive management is involved in too much detail.
- There are no clear guidance and structures in place for operational and tactical management to be able to make decisions.
- There is no formal process that allows for the bypassing of people or structure to expedite decisions or activities.

We suggest the use of the engagement model shown in Figure 8.6. The Strategic Process Council and Chief Process Officer oversee the entire model, with clear lines of relationships and responsibilities shown at each of the three levels.

It can be seen that the process executives and the CBI manager are responsible for the alignment of BPI activities with the organizational strategy and the overall working of the engagement model. They also are responsible for ensuring that the strategy process, CBI and IT are aligned as well as the overall governance, including the monitoring of the progress and impact of business process initiatives and the performance of the processes.

The business managers, process stewards and business process consultants work closely to ensure that a portfolio of improvement initiatives is developed and is progressing so that the overall business processes improve. We have referred to this as tactical management, meaning the planning, funding and organizing of programmes of business process initiatives. The use of the word 'tactical' does not mean it is a temporary or short-term solution. This is the level at which the process architecture is developed, applied and updated.

The staff in the CBI team needs to work closely with their business counterparts (operations staff) on the individual improvement initiatives/projects, as well as the related reporting, analysis and process modeling and documentation.

This tiered structure ensures that there are clear roles, responsibilities and lines of communication between the stakeholders.

In summary, achieving significant and sustainable corporate success with BPI projects and process performance management requires more than just having a successful BPI project or programme: it requires commitment from executive management to make the necessary institutional changes. The two main changes required are

1. The executive needs to manage the business unit manager on the efficiency, effectiveness and robustness of their business processes.
2. The CBI group needs to facilitate the business units to perform better and demonstrate improvements to the executive management.

A CBI engagement model must be introduced to ensure there is the discipline or structure available for all business process management issues to be addressed at the right level, without too many unnecessary escalations to more senior management.

Chapter 9

Project execution

Introduction

Few people would argue that for an organization to have a strategy is a good thing. Few people would also argue that having a strategy is not very helpful to an organization unless it is able to be executed and executed well. Yet, organizations generally find executing strategy *effectively* a very difficult activity. Recently we discussed this aspect with the Dean of Business and IT at a major North American university. The Dean stated that when he meets with senior executives (CxOs) within large organizations, they all have well-defined strategies and *they are all very similar.* In the Dean's view, the difference with organizations that are successful is the ability to execute its strategy exceptionally well.

Kendall (unpublished) argues that 'the fate of any organization is directly linked to its ability to define and rapidly execute the correct strategies to improve. Projects are the primary vehicle for implementing strategic changes. If the senior management team either chooses the wrong projects, the wrong scope or implements too slowly, the organization fails to meet its goals.' Figure 9.1 shows how process execution fits in with the other dimensions.

Why is project execution important?

If, as suggested, strategic objectives can be executed by projects, it follows that project management, the management of projects and the ability to execute them well, is an extremely important capability for an organization to have, and yet even basic project management execution has proven to be a difficult capability and skill set for many organizations to gain at an institutional level. Even though effective project execution is a significant challenge for many organizations, executives continually over-commit organizations to projects beyond the organizations capacity.

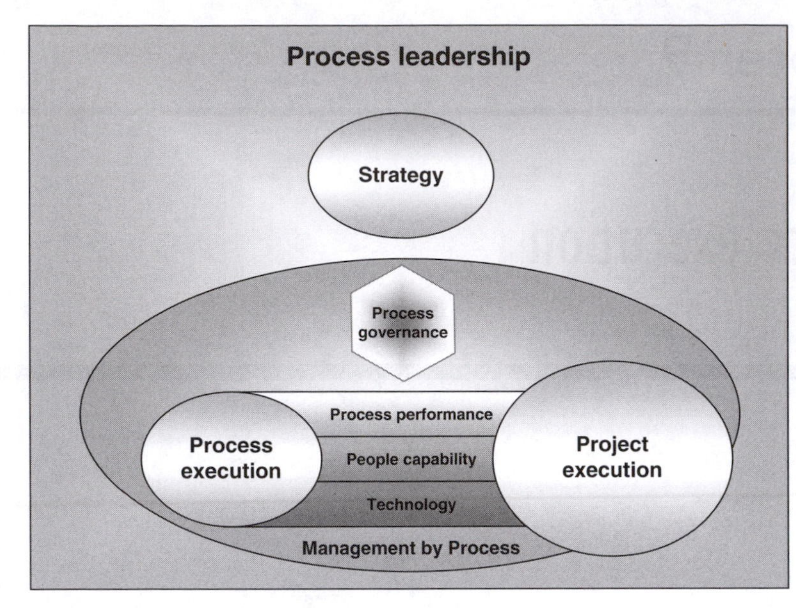

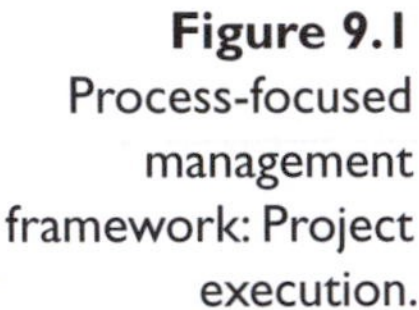

Figure 9.1 Process-focused management framework: Project execution.

There is a large postal organization where 80% of project managers are external contractors. Every time a contractor leaves they take with them the capability and skills needed by the organization to complete projects, and this is not an unusual situation for organizations to find themselves in. Within the same organization, 87% of the current projects being undertaken do not align with its organizational strategy. Why are they doing these projects, especially when the average project length is 3.6 years? There is another large energy utility where the use of its elementary project management methodology is *optional* – 'it is just up to the project manager as to whether or not they wish to use it'!

There are very few organizations that track the business benefits outlined in the original approved business case through to realization and yet many other organizations do not require, or always require, a business case at all. There are still other organizations that do not cost internal resources on projects thus distorting business cases.

It is no wonder that projects fail. A study by Standish Group Inter-national of over 13,000 projects 'shows 82% of all projects surveyed finished late' (Kendall, 2003).

We need to reiterate that the critical link should be: strategy leads to execution via projects However, projects alone do not fulfil strategy – they create the environment and ability for the organization to achieve the strategic objective as a result of the business processes.

If this link is not maintained then projects must not be executed, as discussed in Chapter 7. Put simply, if a project does not contribute towards the realization of one or more strategic objective, then it should not be executed.

There is often an argument put forward by management that a particular project is a 'short term' tactical project that is required to keep the business functioning while we execute the strategy. We have seen organizations implement 'tactical' application systems that are still functioning in the business 27 years later. Once a project, and particularly a business application system, has been successfully implemented into the business, management is often distracted by more pressing business problems and they rarely have the time and resources to replace the 'tactical' solution.

Key trends

There is an ever-growing need for organizations to build not just internal capacity and skills for the execution of projects, but also the internal support structures to manage these projects. Traditionally organizations have build structures such as a programme management office (PMO) to support the management of projects.

The establishment of a PMO has waxed and waned over the years and has taken many forms. Traditionally the PMO has taken one, or a combination, of the following three forms:

1. A governance, control and reporting structure for all projects within the organization.
2. Develop and promote the project management body of knowledge within the organization. This includes not only the adoption or creation of a project management methodology, but also tools, techniques and templates specific to the organization.
3. Provide a group of skilled resources to lead and manage projects for the organization.

The challenge for the PMO is to show value and a contribution to the organization, otherwise funding will not continue to be provided. This is one of the reasons that many PMOs have not survived within organizations. We will discuss this in more detail later in this chapter.

Of recent times, there is a growing trend towards the creation of an internal group with process capability and skills – some have referred to it as a Process Center of Excellence (PCoE). Many of the activities that the PCoE may perform are similar to the PMO. The PCoE can provide a methodology, tools, techniques, templates and pool of skilled resources. The questions that perhaps should be asked: Is a PMO and PCoE different? Should they be the same structure or group? We will address these questions later in this chapter also.

If a PMO or PCoE fails or is not supported (or funded) by senior management, the organization needs to determine why they have not performed and not just cease funding and close them down. Is it poor execution or management within the group? Is it a lack of capability or skills? Is it lack of sufficient funds to enable it to complete its tasks to the level required? Is it lazy management either within the group or at the executive level? By lazy management we are referring to the lack of effort by management to clearly define the roles and responsibilities of the groups, to clarify the needs of the organization for this support structure, staff it appropriately, build the skills and provide adequate funding, and also ensure that the benefits to the organization for the establishment of the groups are realized in practice.

Key elements of project execution

While we do not propose to provide a detailed discussion on how to establish a PMO or the best methodology for managing projects, we will discuss four key aspects of managing an organization from a process-focused perspective. These four aspects are:

1. Selecting the right projects.
2. Setting the projects up for success.
3. Reviewing the trend towards a Process Centre of Excellence and its impact upon an organization's PMO.
4. Standard project management methodologies, such as PRINCE2 and PMBOK, and whether they are sufficient in a process-focused organization.

The reason for discussing these aspects is that, in our experience, they are generally not executed well within many organizations.

Visionary project execution

The visionary position of an organization for this has already been outlined, to some extent, in Chapters 5–7. In the visionary state an organization will only initiate projects that are totally in alignment with the organizations strategy. It will also only initiate the 'right' projects that add the most value to the strategy and the organization, within the constraints of the available investment funds. These projects will have been determined by the governance structure within the organization. The organization will have:

- Two groups involved in the selection of the 'right' projects. The first will review all projects from a 'cost' perspective. It will ensure that the cost side of the business case is robust and capable of being achieved. The second group will be the Strategy Process Council (some organizations refer to this as a Management Board) who will ensure that the benefits outlined in the business case are robust and in balance with the scrutinized costs. Projects are then prioritized based upon the other business priorities and the available investment funds.
- A significant body of knowledge relating to project execution will have been developed for, or enhanced to meet, the specific needs of the organization. This body of knowledge will be continually developed and enhanced as new intellectual property comes to light. The body of knowledge includes a structured approach to the execution and control of projects, programmes and portfolios.
- Significant internal capacity (pool of resources), capability and skill sets will be available to enable projects to be executed on time, budget, to a high quality, *and* always deliver the business benefits outlined in the projects business case. Most of the skills will reside within

the business itself. There is a structured programme to disseminate this expertise and experience throughout the organization.

- Project success is linked not only to the realization of the business benefits outlined in the business case, but to the overall success of the organization.
- A structure is in place that supports the control, reporting and governance of all projects, programmes and portfolio of work within the organization. The governance is appropriate for the size, risks and impact on the project.
- All projects are process-based. That is, the projects understand and pay significant attention to the business process aspects of all projects. Even in technology projects, the business process takes the lead.

Roadmap to project execution

The four areas we will cover in this section are shown in Figure 9.2.

In order to create the best environment for the execution of an organization's projects and maximize the likelihood of success, we would recommend the following four steps:

1. Have a robust and appropriate mechanism for selecting the 'right' projects for your organization.
2. Set them up for success by ensuring that project performance is linked to management, project manager and staff performance, goals and rewards.
3. Ensure that there are mechanisms in place to manage, control and support the projects, such as Portfolio/Programme/Project Management Offices and the Centre of Business Innovation (CBI) as discussed in Chapter 8.

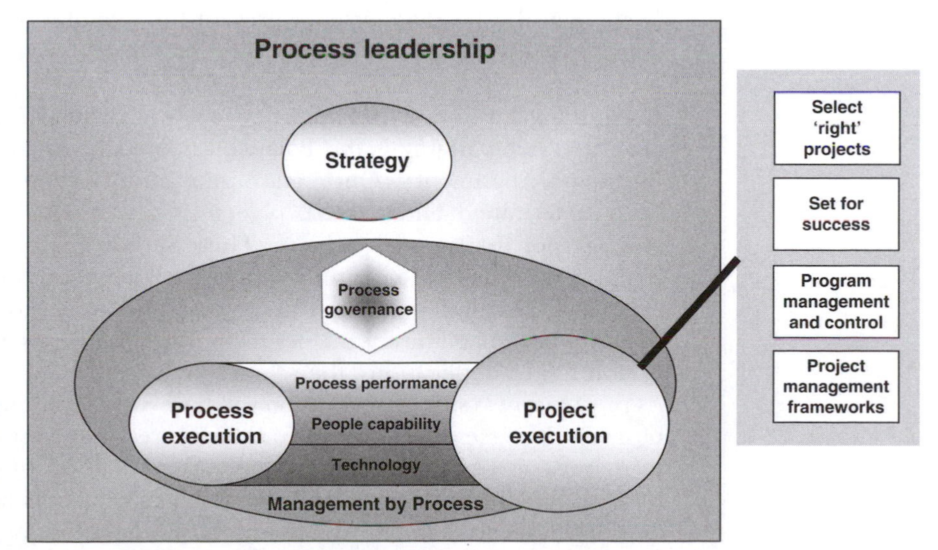

Figure 9.2 Management by Process framework: Project execution steps.

4 Project management methodologies and frameworks are established that will minimize project risk and maximize success.

We will now examine each of these in more detail.

Selecting the 'right' projects

Ensuring organizational strategy success starts with the selection of the 'right' projects to complete. With the exception of IT infrastructure and similar projects, it is usually the business that initiates projects. This starts with an idea or a necessity from the external environment (legislation, business or market imperative), a business case is completed and then the project commenced. Unfortunately, business management often initiates too many projects simultaneously, irrespective of the capacity or maturity of the organization to complete them. We have worked with organizations that have literally hundreds of projects underway. Management claim 'they are doing this for a logical reason – if they do not meet their goals by the next review period, they may no longer be employed or they may miss a significant measurement. Some executives assume that the sooner the project is initiated, the sooner it will be completed (Kendall, 2003).'

While most organizations have a mechanism for the completion of project prioritization, this usually is just a ranking exercise and is not timelined taking into account organizational capacity and capability to deliver the projects or inter-project dependencies. This prioritization exercise also often fails to take account of any other constraints surrounding organizational capacity, such as organizational maturity, ability to, or magnitude of, change.

Figure 9.3 is an example of a simple, yet powerful method of how projects may be 'mapped' to demonstrate to management the impact of a project on the organization. The vertical and horizontal axis may be labeled differently depending upon the message or information you wish to communicate. Each of the numbers shown represents an individual project. The size of the circle may also be changed to represent the size of a project if this is meaningful to the audience. Colors can be added to highlight risk or likelihood of success.

Figure 9.4 shows a pathway that a strategically motivated project may take; another pathway is the Balanced Scorecard as outlined in Chapter 7 (Strategic Alignment). Once an organization's strategy (objectives) have been determined and agreed, objectives are usually allocated to one or more senior managers to achieve. These senior managers then create their initial action plans of how they expect to achieve each objective. When the options are examined, management must take into account the capacity and operational capabilities of the organization to fully execute the option. If the option has IT implications, then the IT architecture must also be considered. As most business strategy execution will have an impact upon the organizations business processes, the enterprise architecture, or at least the process architecture, must also be taken into account. As a result of the various options put forward, the final option will be selected, project prioritized (e.g. Figure 9.3) and then executed, to deliver the organizational strategy.

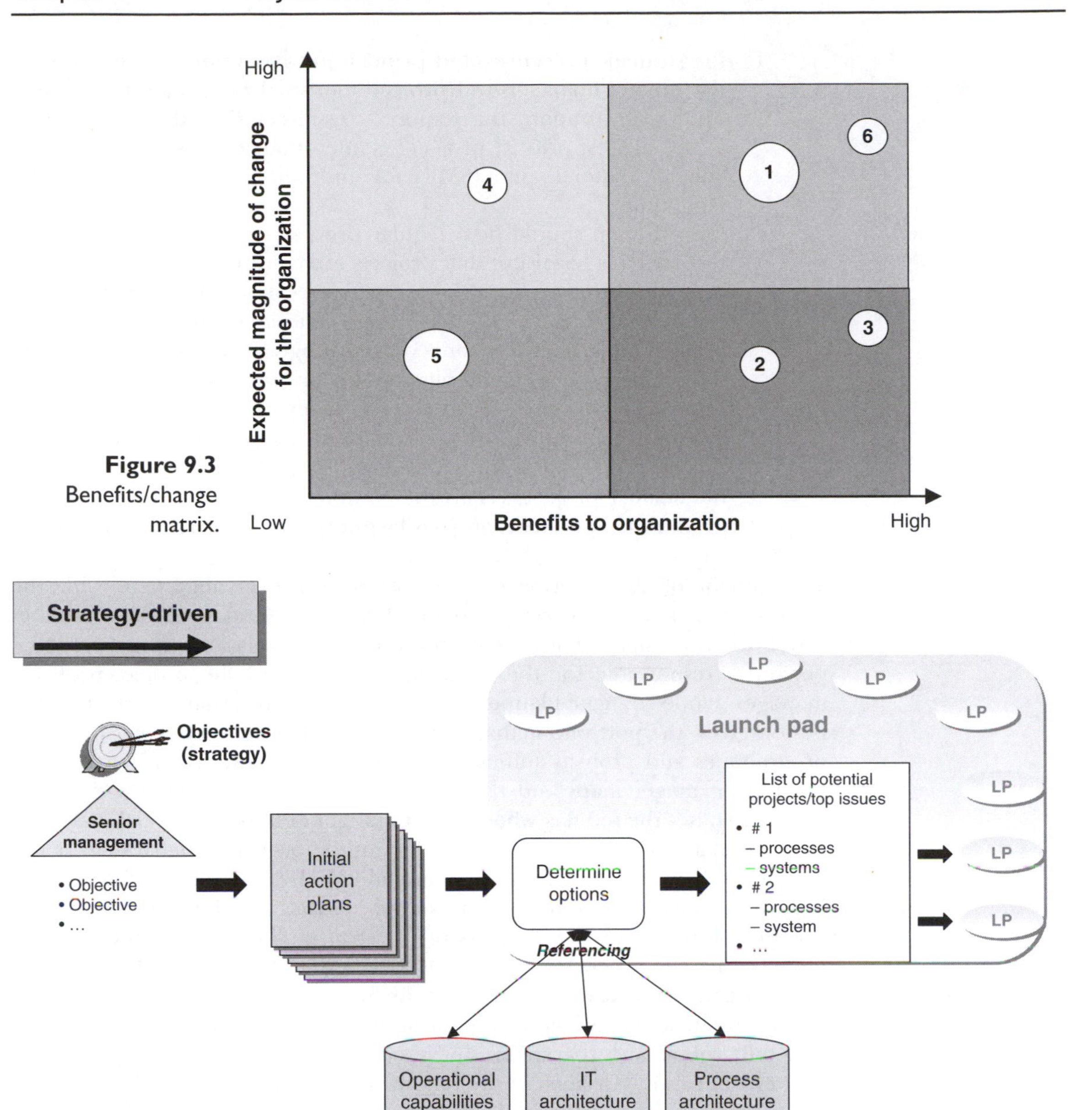

Figure 9.3
Benefits/change matrix.

Figure 9.4
Strategy-driven projects.

Prioritization requires more than just an occasional or casual review of project priorities. It requires discipline in at least the following areas:

- There can only be a single priority 1 project – there are still organizations that have numerous priority 1 projects. In fact, some have so many priority 1 projects that they are unable to complete them – let alone other priorities.
- Projects with low priorities should not take away resources from higher priority projects. Resource allocation should be based on

the strategic objectives and priorities rather than on the ability of the project manager and project sponsor have projects approved. After prioritization, the required resources should be committed to the highest priority projects. Some organizations book the same Subject Matter Expert (SME) for multiple full-time projects at the same time!

- Prioritization should be a regular process – that is conducted on a regular basis to ensure that projects are continually aligned with the strategic objectives. In addition, every time a new project is requested it should be ranked. This mechanism can reduce the initiation of low priority projects as it is ranked accordingly at the start. Pragmatism should be used when implementing prioritization: it is not just a case of placing projects in reverse order of priority and assigning resources to top priority projects until the resources run out. The process should allow for exceptions, for example, the completion of five smaller projects versus the completion of one large project; or deadlines imposed on projects by internal or external authorities.

The benefit of this approach is that projects are in alignment with the organizations strategy, as discussed in Chapter 7. Kendall (2003) describes it well when he states that 'usually, there is a close relationship between the person(s) responsible for the strategic planning and the project portfolio manager. While strategic planners identify the ideas necessary to meet organization goals, the portfolio manager makes sure that there are corresponding programmes and projects sufficient to accomplish those ideas. Further, the portfolio manager maps and tracks the project execution against the strategies and raises the red flag when there is danger of missing a goal. Finally, the portfolio manager also lets strategic planning know when the strategy is not practical relative to project resources available.' The emergence of a Strategic Portfolio Management Office is an example of this. Another way of ensuring executive support for the progress of the strategic objectives is the appointment of 'theme' owners for the themes specified in the Strategy Map.

In Chapter 5 we described how it is the Strategic Process Council's responsibility to select and prioritize organizational projects. However, it is critical that only robust and realistic business cases are submitted for approval. An example of this is described in the following case study.

Case study: Robust business cases

We were recently discussing business cases and projects with a senior executive in a large bank. He described how one of the bank's project managers was recently blamed for the 'blow-out' of the cost of a project when it went from $5 to $15 million. However, in reality it was not the fault of the project manager. The executive described that the project was never a $5 million project and that either the business case costs submitted were minimized to 'get the project approved' or the costs were poorly estimated in the first place.

He then described the opposite situation where the project manager was seen as a hero for bringing a project substantially under-budget ($10 million from the budgeted $20 million). It was always a $10 million project – the costs were poorly estimated in the first place.

(Continued)

Case study: Robust business cases (*Continued*)

Message: These examples highlight the need for an independent group within an organization to challenge the cost side of all business cases. This is ideally completed before a business case is submitted to the Strategic Process Council for approval and prioritization. Also there needs to be 'feedback' to the managers signing-off on the business case costs with regard to their accuracy.

As stated earlier, the Strategic Process Council must not only prioritize and approve business cases and projects, but also review and ensure that the benefits are achievable and able to be realized.

Set for success

Within some organizations the performance and worth of executives is judged by the number of projects they initiate rather than successful completion; and some managers are extremely adept at hiding project performance (or lack of performance). A similar situation can arise with some programme or portfolio management offices (PMO).

Setting for success is about establishing a Project Performance Framework. This is achieved by creating a set of criteria, measures and performance links from project outcomes to various project responsibility levels within the organization.

We will use the PRINCE2 project management methodology as an example of the layers of responsibility within a project and relate this to our governance structure and project framework. PRINCE2 suggests that a Project Board be established as shown in Figure 9.5.

In the Office of Government Commerce (OGC, 2002, p. 201) publication on PRINCE2, it is stated that 'the Project Board is not a democracy controlled by votes. The executive is the key decision-maker with advice and commitments from others.'

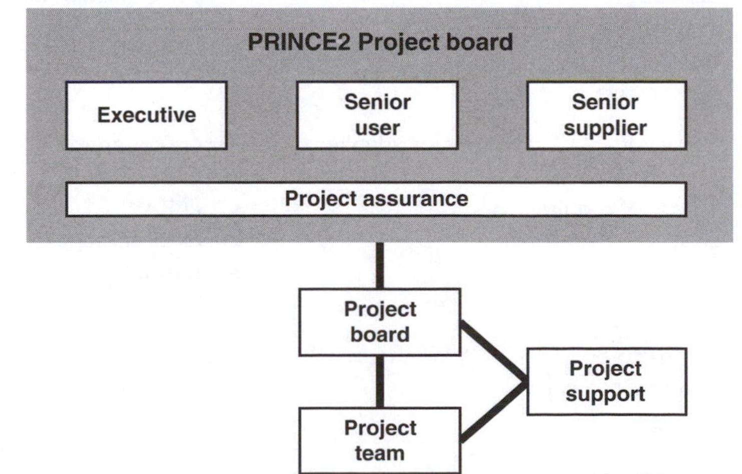

Figure 9.5 PRINCE2 project management structure (OGC, 2002).

In Table 9.1 we will compare the responsibilities, as outlined in PRINCE2, with our suggested process governance structure and management by process approach. We will also suggest what each level of management should be responsible for within a project.

We would like to highlight several issues associated with Table 9.1.

1 Project funding will sometimes be separated from the responsibility realizing the project business benefits. Some organizations will have a business sponsor and a business owner, with the sponsor being the person who provides the funding and takes overall responsibility for

Table 9.1
Project board roles and responsibilities

PRINCE2		**Management by Process Framework**	
Role	**Delivery responsibilities**	**Role**	**Delivery responsibilities**
Executive	• Business needs (project scope) • Value for money	Process executive	• Business needs as defined in the "Red Wine Test" (project scope is only one part of this) • Ultimate responsibility for the delivery of project success and business benefits as outlined in business case. These are usually delegated to the process steward, however, the process executive has the ultimate accountability.
Senior user	• Use of products (project deliveries) • Achieving project objectives • Delivery of user benefits • Project outcomes from a users perspective	Process steward	• Approving definition of business requirements • Actual delivery of business benefits as outlined in business case • Smooth transition to sustainable performance (business as usual)
Senior supplier	• Supply of resources • Supply of required skills	Senior supplier	Resources and skills may be supplied by: • CBI • PMO • External suppliers • Business • IT
Project assurance	• Needs to be external to project manager	PMO and/or CBI	Project success will be supported by: • Training • Coaching • Project quality reviews

the project and the business owner is the equivalent of the senior user/process steward and responsible for the delivery of the business benefits. In the initial stages of a project, especially a process improvement project, the funding should revolve around 'gates' and the delivery of the project phases described in the later section on Project Management Frameworks.

Process executive and process steward's targets (Key Performance Indicators (KPIs)), performance and rewards must be clearly linked to the success of the projects they are responsible for. The best way to ensure delivery of the business benefits outlined in the business case is, once a project has been approved, to determine the date that the business benefits are targeted to commence accruing and delete the amount from the business budget. This will specifically focus the business managers to ensure benefits are delivered.

2 The process steward should have delivery responsibility for all items mentioned in Table 9.1 from both the PRINCE2 and Management by Process Framework columns. Projects are not initiated to satisfy 'users' (by the way, they should be called 'customers', because they can disengage with you – it is called outsourcing), projects are initiated to provide added value and increased business benefit to the organization. The performance links for process stewards are outlined in Chapter 5.

3 The senior supplier can be internal or external. The internal groups that can add significant value to the project include:

- CBI – who could provide resources, coaching, training and advice to the project (Chapter 8 describes the function of the CBI);
- the PMO could also provide similar services to the CBI;
- the business must also understand that they have a significant responsibility of delivering SMEs and other information to the project;
- if there is an IT component, then IT will be required to provide resources and expertise.

Individual members of these groups must have their personal performance targets, assessments and rewards linked to the success of the organizations projects. When project success and performance are linked a change of attitude is created that moves these groups (CBI and PMO) from 'all care and no responsibility' to an attitude of coach, accountability and responsibility – the PMO/CBI only wins if the project is successful.

4 Project assurance must never be delegated to the project manager. Often it is considered to be the role of the Chief Financial Officer, but they are rarely close enough to the project to provide any real input or benefit. The best placed part of the organization will be the PMO and the CBI. These groups must minimize the risks associated with projects by providing training and coaching at a generic level, and then conduct specific project quality reviews at a detailed level.

While PMOs are considered, by most organizations, to be valuable because projects are valuable, they are often perceived to only have a reporting or monitoring role with little accountability, other than reporting, or being the keepers of the project management methodology. PMOs need to have an appropriate level of ownership, accountability and responsibility for project outcomes and should have its success clearly linked to the success of projects. Individual members of the PMO must have their personal performance targets, assessments and rewards linked to the success of the organizations projects. This will provide a focus for the PMO team to deliver their services.

Programme management and control

In our previous book, Jeston and Nelis (2008), we provided a framework for the successful and repeatable implementation of process improvement (business process management (BPM)) projects. However, there is a significant difference between the successful implementation of individual BPM projects and the 'management of projects' across a business unit or organization. We have entitled this book *Management by Process* because we believe that 'managing by processes' is a management philosophy that organizations must adopt to assist in organizational success and this can be applied to the management and control of projects or programmes also.

A PMO can represent different entities within an organization. It could be considered either a Portfolio or Programme or Project Management Office. For the purposes of this discussion we consider all three and we will simply refer to it as a PMO.

Types of PMO

PMOs can cover a number of activities and an organization needs to determine what it wishes to achieve with this department. The functions and types could cover all or some of these:

- *Strategic or enterprise PMO* – where the PMO assists executives in the portfolio management necessary to achieve the outlined strategic objectives. The scope of this PMO will relate to the entire organization.
- *Weather station* – where the PMO simply reports upon the status of projects (budgets, timelines, issues, risks, actual progress).
- *Control tower* – where the PMO establishes a set of standards (project management methodology) that must be adhered to; provides advice, enforcement (audits) and continuous improvement for these standards.
- *Resource pool* – the PMO maintains a pool of trained project managers who are 'loaned' out to the business to complete projects.
- *Coaching and training* – PMO provides coaching and training for the project managers and project team. This implies that the coach also has some responsibility for the performance of the people they are coaching.
- *OAR* – there is genuine Ownership, Accountability and Responsibility (OAR) for the outcomes of projects. This means that the PMO/CBI has targets and rewards that are linked to the success of project outcomes.

There is no right answer for an organization. Different organizations will have different needs, depending upon where they are now and how it is growing. The maturity of the organization with regard to project management will have a large impact upon the structure, responsibilities and expectations for a PMO.

Let us spend some time looking at one of the project management maturity models to enable us to place this in perspective.

The biggest challenge for most PMOs and CBIs is to show value to the organization. Initially it might be better to focus on some key projects and processes, achieve significant results and start growing from there, rather than forcing all projects and processes to make an immediate use of the PMO and/or CBI. Some executive managers will hopefully notice and appreciate the services and start demanding similar approaches and disciplines from all project managers and process stewards.

Project management maturity models

The high percentage of project failures that we have mentioned previously is even more alarming and remarkable when we take into account the amount of project management literature, training and expertise that has been available over the last 15 years. Project management should no longer be considered an art that is only known and shared with the select few outstanding project managers. With the considerable growth in project documentation and the well-respected methodologies and training from the various project management institutes, projects success should have become much more repeatable and considered more of a science. However, it has not, and while it is getting better, it is still nowhere nearly good enough. We have found that one of the root causes of the low level of project success is that organizations do not have the required understanding, discipline and maturity to make the right decisions.

Case study: More than project managers are required

A multinational service organization was struggling with several strategic projects that were of high importance and lacking progress. Management brought in an experienced project management company and asked them to successfully manage their key projects and provide coaching and mentoring services to its internal project managers. The vendor project managers noticed that there were no project documents such as a project plans, schedules and budgets. The vendor project managers started preparing these documents but soon realized that the key problem lay in the lack of understanding and commitment from the organization. The lack of progress in several of the projects was escalated through 'yellow' and 'red' status. The executives complained that the external vendor project managers were incompetent as they thought that a status of yellow or red was suggesting that the projects be stopped.

Message: Successful projects require more than just experienced project managers. They also require a disciplined organization that understands the importance of all the aspects of project management and is willing and able to make timely and well-founded decisions.

We suggest that there are the following misconceptions regarding project management:

- *Project management maturity only relates to the project managers themselves*
  Maturity relates to all relevant stakeholders within the organization, such as project sponsors, all project stakeholders and the project steering committee.
- *Project management maturity relates to an organization's existing project processes, templates and documentation*
  Maturity relates to how the organization executes, manages and governs projects. It is about the actual usage of the project processes and templates. We have seen many organizations that strive to have excellent project processes and document templates, and forget to focus on the actual deployment and mandatory usage of these processes and documents.
- *Determination of project management maturity level based on vague statements*
  Each of the maturity levels has a brief description associated with it, such as Optimized Process, Managed Process, Defined Process, Repeatable Process and Initial Process. Unfortunately, some organizations use this brief description to determine and describe their maturity level. Whereas, there are multiple levels in between and the maturity level may vary for different parts of the organization. This simplistic subjective assessment of maturity needs to be more factually based using a more formal and defined measuring process – especially as most organizations overestimate their current maturity.
- *Maturity can be externally purchased or implemented instantly*
  As maturity relates to how an organization as a whole understands and performs project management it is impossible to raise it through various levels in maturity within just a few months. External consultants and project manager can provide assistance, but cannot replace internal capability and capacity.

OGC's maturity model

This model (OGC, 2006) was developed specifically for project, programme and portfolio management and the distinction between these three is especially powerful. The model provides significant details for each of the processes specified (Table 9.2).

OGC has a process-focused approach to project management and maps the specific project management processes to specific maturity levels (Table 9.3).

If an organization gains a better understanding of its level of project management maturity this will assist it in determining how to improve and move forward to higher levels. Does the organization need to:

- improve its project templates, documentation or project management methodology;
- build better project management capability and skills;
- improve the level of executive management decision-making and management of projects.

There is a need for the organization to determine what is its required/desired level of maturity for the organization.

Table 9.2
OGCs maturity model (OGC, 2006, pp. 7–8)

Maturity	Project	Program	Portfolio
Level 1: *Initial process*	Does the organization recognize projects and run them differently from its ongoing business?	Does the organization recognize programmes and run them differently to projects?	Does the organization's Board recognize programmes and projects and run an informal list of its investments in programmes and projects?
Level 2: *Repeatable process*	Does the organization ensure that each project is run with its own processes and procedures to a minimum specified standard?	Does the organization ensure that each programme is run with its own processes and procedures to a minimum specified standard?	Does the organization ensure that each programme and/or project in its portfolio is run with its own processes and procedures to a minimum specified standard?
Level 3: *Defined process*	Does the organization have its own centrally controlled project processes, *and* can individual projects flex within these processes to suit the particular project?	Does the organization have its own centrally controlled programme processes *and* can individual programmes flex within these processes to suit the particular programme?	Does the organization have its own centrally controlled programme and project processes *and* can individual programmes and projects flex within these processes to suit particular programmes and/or projects. And does the organization have its own portfolio management process?

(Continued)

Table 9.2 (*Continued*)

Maturity	Project	Program	Portfolio
Level 4: *Managed process*	Does the organization obtain and retain specific measurements on its project management performance *and* run a quality management organization to better predict future performance?	Does the organization obtain and retain specific measurements on its programme management performance *and* run a quality management organization to better predict future programme outcomes?	Does the organization obtain and retain specific management metrics on its whole portfolio of programmes and projects as a means of predicting future performance? Does the organization assess its capacity to manage programmes and projects and prioritize them accordingly?
Level 5: *Optimized process*	Does the organization run continuous process improvement *with* proactive problem and technology management for projects in order to improve its ability to depict performance over time and optimize processes?	Does the organization run continuous process improvement *with* proactive problem and technology management for programmes in order to improve its ability to depict performance over time and optimize processes?	Does the organization run continuous process improvement *with* proactive problem and technology management for the portfolio in order to improve its ability to depict performance over time and optimize processes?

Table 9.3
Processes in each maturity level (OGC, 2006, p. 9)

Level 1: Initial process	
1.1	Project definition
1.2	Programme management awareness
Level 2: Repeatable process	
2.1	Business case development
2.2	Programme organization
2.3	Programme definition
2.4	Project establishment
2.5	Project planning, monitoring and control
2.6	Stakeholder management and communications
2.7	Requirements management
2.8	Risk management
2.9	Configuration management
	Programme planning and control
	Management of suppliers and external parties
Level 3: Defined process	
3.1	Benefits management
3.2	Transition management
3.3	Information management
3.4	Organizational focus
3.5	Process definition
3.6	Training, skills and competency development
3.7	Integrated management and reporting
3.8	Life cycle control
3.9	Inter-group co-ordination and networking
	Quality assurance
	Centre of Excellence (COE) role deployment
	Organization portfolio establishment
Level 4: Managed process	
4.1	Management metrics
4.2	Quality management
4.3	Organizational cultural growth
4.4	Capacity management
Level 5: Optimized process	
5.1	Proactive problem management
5.2	Technology management
5.3	Continuous process improvement

However, we can make some generalizations about growing an organizations project maturity:

- *Process leads execution* – the processes need to be developed for the next maturity level prior to training and deployment.
- *Realistic timelines* – building maturity is not achieved overnight, it needs an enormous amount of effort, dedication and communication to increase in maturity level.
- *Right sequence of processes* – as the higher levels of maturity rely on the presence and correct execution of the underlying processes.
- The adherence to the process needs to be measured in a pragmatic way without too much additional overhead. Much of the information can be obtained if using a centralized project repository. A qualitative assessment should also be completed of the project management documentation and activities.

PMO compared to a CBI

As mentioned earlier, most organizations have some type of PMO and with the growing awareness of the importance of business processes and BPM some organizations are now starting to create centres of knowledge and expertise around its business processes. Many organizations refer to these groups of BPM Centres of Excellence. We would prefer to call them a Centre of Business (Process) Innovation (CBI). As mentioned previously, a PMO may fulfil many roles within an organization. The role within each organization will depend upon several factors: the maturity of the organization, the organizational needs and problems, what governance is in place, and skills and capabilities of the organization's project managers and business sponsors. The left-hand column in Table 9.4 is headed 'PMO' and shows the typical roles and benefits that a PMO may bring to an organization. The right-hand column shows the roles and benefits to be gained from a CBI. It is interesting to compare the two.

Both a PMO and a CBI may bring significant benefit to an organization. In addition to the benefits indicated in Table 9.4, they could include (Goodpasture, 2000):

- Professionalism, organizational competency and capability maturity.
- Re-use of organizational knowledge and investment in future projects, which can lead to lower costs going forward.
- Mitigation of risks and avoidance of mistakes.
- Improved resource allocation across projects and within portfolios of projects.

The question arises, should these two organizational activities be one? Once again, there is no easy answer to this question. It will depend upon the maturity of the organization, the people within it, the acceptance of a project management methodology, the PMO, and the understanding of the importance of business processes. All things being equal, which they never are, then we believe that there is an opportunity of merging these two groups into the one group. As seen in Table 9.4, there is significant overlap in functionality and responsibilities. Both have successful projects as a central and critical theme for the measurement of its success and adding benefit to an organization.

Table 9.4
PMO and CBI roles and responsibilities

	Programme Management Office (PMO)	Center of Business Innovation (CBI)
1	Develop body of knowledge (intellectual practical) of project and programme management. This will include the provision and continual development of templates and tools.	Develop and continually enhance the body of knowledge (intellectual practical) of process improvement projects and management by process. This will include the provision and continual development of templates and tools.
2	Provide project coaching and mentoring to project managers and portfolio coaching and mentoring to line managers.	Provide coaching and mentoring for process executives, process stewards, project managers, process analysts and project team members.
3	Be a center of expertise with regard to project management.	Be a center of expertise with regard to project management of business process improvement projects.
4		Be a center of expertise with regard to the management of business processes and their continual improvement.
5	Provide a pool of project management resources.	Provide a pool of resources covering business process improvement: project managers, process analysts and other process experts and could include business analysts for core Business Process Management Systems.
6	Organizational reporting of all projects within the organization, including: status, quality, projects at risk.	Organizational quality assurance and audits of all business process improvement projects.
7	Governance of projects across the organization.	Ensuring that the process governance structures, roles, responsibilities, targets (KPIs and KRAs) are in place and appropriate for various areas of the organization.
8		Develop and maintain the foundation business process principles.
9		Develop and maintain the process architecture.

However, the amalgamation probably needs to evolve over time and not be 'pushed' into existence by management. If an organization just 'does not yet get processes' then pushing it into existence will potentially have a detrimental impact upon the process journey and process maturity within the organization.

Project management frameworks

In our experience of engaging with many businesses in the journey of becoming a process-focused organization and the development and implementation of process improvement projects, we have learned that appropriate project management is critical to risk minimization and successful, repeatable projects. The key word in the previous sentence is 'appropriate'.

Does this mean the normal execution of PRINCE2 or PMBOK project management methodologies?

The answer to this question is predominantly 'no'. PRINCE2 and PMBOK definitely have their role and make a significant contribution to project success, but they are simply not enough – more is needed.

In the journey towards becoming a more mature process-focused organization there will come a time, and probably early on, when an organization must review its current project management methodology to determine how appropriate it is for the organization that it wishes to become. This could be completed as part of the CBI activities.

As stated, in our experience, more is needed than these standard project management methodologies provide. Experienced project and programme managers will have their own ways of overcoming the deficiencies of these methodologies, but an organization must not become reliant upon the specific skills and experience of individual project managers. This is especially the case for the many organizations that predominantly rely of external contractors and consultants to provide the bulk of these skills. The organization must add to these project methodologies and make them suit its needs.

Before we examine what is missing from these methodologies let us look at some of the traditional project challenges; things such as project delays, budget overruns, scope creep and changing requirements, not enough resources, inconsistent project management practices across the organization, lack of planning, unclear priorities, poor communications, always re-inventing the wheel.

The typical running of projects is illustrated, in a somewhat provocative way, by the 15 immutable laws of project management (www.ifaq.wap.org/science/lawprojman.html accessed 5 September 2007):

- **Law 1:** *'No major project is ever completed on time, within budget, with the same staff that started it, nor does the project do what it is supposed to do. It is highly unlikely that yours will be the first.'*
  Not only is this appalling, but even if you somehow managed to deliver on time, budget and to a high level of quality, there is no guarantee that you will deliver what the business needs.
- **Law 2:** *'One advantage of fussy project objectives is that they let you avoid embarrassment in estimating the corresponding costs.'*
- **Law 3:** *'The effort required to correct a project that is off course increases geometrically with time.'*
- **Law 4:** *'The project purpose statement you wrote and understand will be seen differently by everyone else.'*

  'Corollary: If you explain the purpose so clearly that no one could possibly misunderstand, someone will.'
- **Law 5:** *'Measurable benefits are real. Intangible benefits are not measurable, thus intangible benefits are not real.'*
- **Law 6:** *'Anyone who can work effectively on a project part-time certainly does not have enough to do now.'*
- **Law 7:** *'The greater the project's technical complexity, the less you need a technician to manage it. Get the best manager you can. The manager will get the technicians.'*

- **Law 8:** *'A carelessly planned project will take three times longer to complete than expected. A carefully planned project will only take twice as long.'*
- **Law 9:** *'When the project is going well, something will go wrong.'*
- **Law 10:** *'Project teams detest weekly progress reporting because it so vividly manifests their lack of progress.'*
- **Law 11:** *'Projects progress rapidly until they are 90% complete. Then they remain 90% complete forever.'*
- **Law 12:** *'If project content is allowed to change freely, the rate of change will exceed the rate of progress.'*
- **Law 13:** *'If the user does not believe in the system, a parallel system will be developed. Neither system will work very well.'*
- **Law 14:** *'Benefits achieved are a function of the thoroughness of the post-audit check.'*

  'Corollary: The prospect of an independent post-audit provides the project team with a powerful incentive to deliver a good system on schedule on budget.'
- **Law 15:** *'No law is immutable.'*

We would now like to discuss and gain some understanding of why most of us can relate to these immutable laws and what can be done to eliminate (or at least minimize) them. We need to fulfil Law 15 and overcome all of these immutable laws.

Project management methodologies are specifically orientated towards delivering according to the 'scope' as defined in the Project Initiation Document. There is an implicit assumption here, and it is that the 'scope' will deliver business benefits. The focus must be towards delivering business results and it is these business results that must be in the forefront of the project teams mind, not the 'scope' itself. Now we know that the project managers reading this will argue that business results and scope are the same thing – all the time. But are they really? The external and internal circumstances change, often in different ways than anticipated. Sometimes the purpose of a project is not revisited as business circumstances change, whereas most projects will require at least some level of change. In other words, business demands are the reason for the existence of projects and changes are not a disruption to the project execution. Hence an open mind to changes is critical and a strict discipline for project change control is essential.

If projects are selected carefully and appropriately, then the next step is to define the projects desired outcomes, which is potentially different to the project scope. We are not saying that a project should not have a scope. There must always be a target that the project team strives towards and knows when it has been achieved. It is just that the definition of the business results/outcomes is difficult and critical, especially for business process improvement and innovation initiatives. So the challenge is how can this be achieved to ensure that the business is satisfied with the project outcomes and yet not have the project direction change so much for the project team that nothing is delivered (thus avoiding the fulfilment of Laws 1, 4, 12 and 13).

The first thing we can do is to complete the 'Red Wine Test' as outlined in Jeston and Nelis (2008, Chapter 15). This is best completed in a workshop setting ensuring that all the significant business stakeholders are in the room at the same time. The facilitator of this workshop must ask the following question, ensuring that he/she creates the necessary emotional imagery to ensure that stakeholders are relaxed and in a state of mind to answer the question.

> It is six weeks after the completion of the project and you are sitting back at home on a cold winters evening in front of a warm fire having a red wine and reflecting on the project. You decide that it has been outstandingly successful. Why?

The answer to this question will cover not only what the project must deliver, but also how it was delivered and the state of the business as a result of the project delivery.

The reasons given during the workshop must cover both project and business outcomes. Projects are not solely delivered by project teams, businesses must be intimately involved and committed.

When a project manager and project business sponsor try to determine the scope of the project before it commences, it is a very difficult activity. They are writing a statement or document that is projecting into the future. They are writing words which describe for the stakeholders that 'if we deliver this, many months or years in the future, the project will meet the following projected businesses needs'. This is an extremely difficult thing to do, and get right.

Whereas, the Red Wine Test is placing the significant stakeholders in a state of mind that places them in the future looking back and asks why the project has been successful? It could be argued that the outcomes from the Red Wine Test are the critical success checklist items, but they are more than this. The list provided by the stakeholders will usually be able to be split into two separate sections: those that can and will be delivered by the project team, and those that cannot. This is also extremely important because it enables the business to understand what activities they are responsible for delivering over and above the project team.

The second aspect is for the project team and business to understand that, especially on process-driven projects, the project scope will not actually be defined until you are well into the project. We believe that this is true of many, or most, business projects. It could be argued that there are in fact several projects, or more preferably one project with multiple project gates/stages. The scope will evolve and change as a result of the outcomes of each stage of the project. For example, the stages could be:

1. *Discovery phase*: Where the business has an idea for a project and this idea needs investigating to determine if there is the opportunity to deliver business benefits.
2. *Launch Pad*: If business benefits are available, then the first part of the project will need to be launched. This includes project planning, resourcing, costing, define the initial scope, Red Wine Test and outcomes for the next stage.
3. *Understand phase*: This next stage would typically include the Understand phase (Jeston and Nelis, 2008, Chapter 16): process modeling the current processes (only in sufficient detail to enable the processes to be redesigned), baselining of the metrics of the current processes, and baselining of the skills and capabilities

required to execute the current processes. Then the scope of the next stage will need to be determined. Metrics are critical to be able to show true value and potential future benefits.

4 *Innovate phase*: Various options are developed for the redesign of the processes, as well as the new metrics, and skills and capabilities. This is where the business requirements will be process modeled and described in detail, which will enable a comparison and decision to be made of what the business requirements are and thus a 'final' scope. The scope will not only include any business application development activities, but also the business training (a comparison between the baseline skills and capability with those required as a result of the redesigned processes). The comparison of the new metrics with the baseline metrics will also provide guidance for the realization of business benefits.

The reader can see how the 'scope' has evolved over the project and into its final form. This evolution has allowed the organization to more thoroughly understand what its business requirements are and build a robust business case by understanding its 'real' requirements. If this business requirement evolution has been handled with significant stakeholder engagement, as it must, then the project risk will have been significantly reduced and the likelihood of a successful project will have been achieved.

Most project managers believe that they have successfully completed a project once it has been implemented and it has 'gone live'. This is absolutely *not* the case. Business projects are commenced because they will add value to the business. This added value will have been outlined and approved in the projects business case. It is within the 'scope' of the project to deliver the benefits designated in the business case. This is referred to as the Realize Value phase in the Jeston and Nelis 7FE Project Framework (2008).

While it is the businesses responsibility to realize the benefits, this can only be achieved with the assistance of the project team who must deliver the project in such as way as to allow this to occur.

It is usually the project business owner/executive, with the assistance of the project manager and process steward, who is responsible for ensuring that the benefits are realized.

Usually as the project benefits are being realized, the project team is transferring 'ownership' of the project to the business, so that it becomes a business as usual activity. We refer to this as the Sustainable Performance phase on our 7FE Project Management Framework (Jeston and Nelis, 2008) (Figure 9.6). The business must take ownership of the project deliverables, ensuring that they can be sustained and continuous business process improvement takes place as part of the business as usual activities, and not continually via projects.

The project activities mentioned previously are not part of the PMBOK or PRINCE2 project management methodologies or they are not covered in sufficient detail. While we are actually very supportive of project management methodologies, they are simply not good or extensive enough in their current state.

Figure 9.6 shows the perfect way for these two project management approaches to relate. The five stages of the PRINCE2 methodology have a clear relationship to the ten phases in the 7FE project framework.

In reality, most projects are completed as shown in Figure 9.7. In our experience this represents how most organizations either do not complete at all

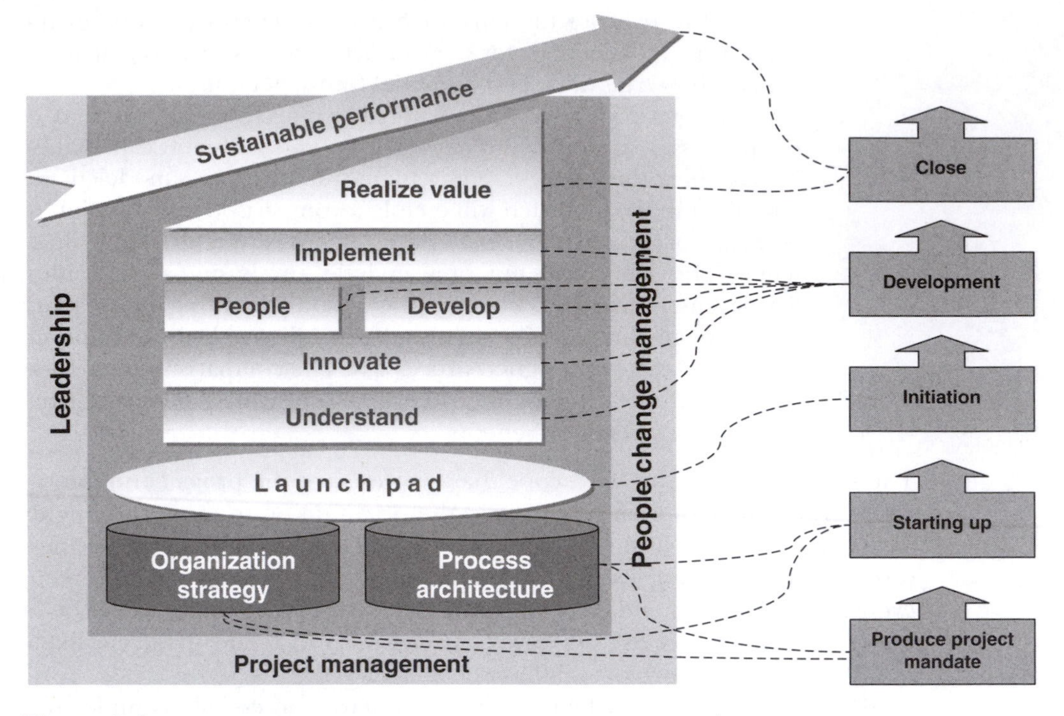

Figure 9.6
7FE Project Framework compared to PRINCE2 methodology.

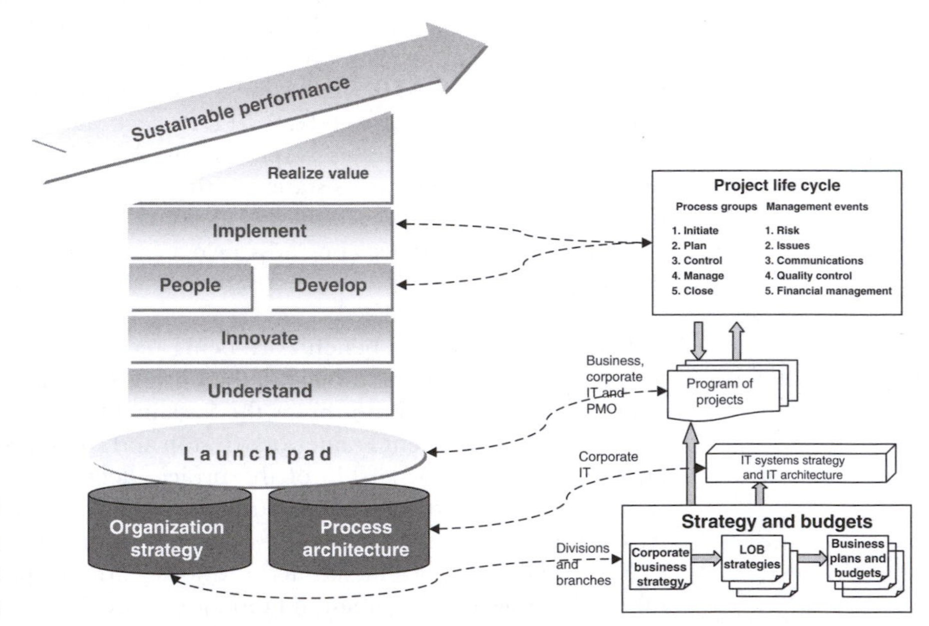

Figure 9.7
7FE Project Framework verses typical project flow.

or at least not well, the Understand, Innovate and People phases of the 7FE Project Framework approach. IT and project managers simply think that it is the businesses problem to solve any impacts upon the business processes and operational staff roles or job functions. When these activities are completed well, they add significant value to the business outcomes of the project implementation. Similarly, most project managers believe that they have completed the project once it is implemented, whereas the benefits realization and sustainable performance phases are also critical, as explained previously.

Figure 9.7 shows the situation in most organizations where:

- There is no process architecture – in fact, most organizations do not truly understand what a process architecture is.
- It is rare for a project to process model the current or future state business processes to ensure that any business application developments integrate with them, or understand how they must change.
- Any changes to the roles and responsibilities of business staff, as a result of the project, are seen as a business issue and nothing to do with the project.
- We have already commented on realizing business value and sustainable performance previously.

Conclusion

For those who insist that PMBOK or PRINCE2 project management methodologies are sufficient and all that is needed for successful projects, then we ask:'why do we have such horrific evidence of project failures?'

The Standish Group International study of over 13,000 projects described at the commencement of this chapter clearly shows that too many projects fail. While the 2006 Standish Group Chaos Report, as yet unreleased (www.sdtimes.com/article/story-20070301-01, accessed 11 September 2007), shows improvement, there is still a long way to go. This report shows that project management, compared to 1994, has improved, but still only 35% of software projects are categorized as successful (completed on time, on budget and met user requirements). The study still 'shows that *only* 19% of projects were outright failures'. *Only – that is one in five fail! That is appalling.*

It then goes on to say that 'projects described as challenged, meaning they had cost or time overruns or did not fully meet the user's needs, declined to 46% in 2006 from 52.7% in 1994'. *46% of projects are challenged and 19% are outright failures* and yet organizations continue to promote the *sole* use of a project management methodology as a cure for their poor record of project success. If projects fail, 'it must have been the fault of the project manager, poor scope, poor business engagement and so on'. It is 'never' the fault of the methodology! While the reasons for failure mentioned previously may well be true in certain circumstances, it is also true that the way organizations currently complete projects is also very poor.

> The really frightening thing for us is that projects can be delivered on time, within budget, to a high standard of quality, having met the project scope, and still not meet the business' expectations, and therefore, be a failure.

In our experience, ensuring that the Red Wine Test, scope evolution, benefits realization, stakeholder management and sustainable performance activities are completed, substantially reduces project risk and significantly increases the chances of success.

Chapter 10

Future of business process management

Looking into the future is always a challenging and difficult thing to do. If we were to take our crystal ball and our experience then we would say that the following trends are either emerging or strengthening within the business process management (BPM) environment.

BPM focuses on business performance management

Executive management will give more and more attention to the business processes within their organization and especially the outcomes of these processes, which of course is the processes performance.

With the increasing demand to continue to improve, performance organizations have exhausted the low-hanging fruit. Blindly cutting staff numbers across the organization is no longer possible without seriously affecting the performance and compliance of the organization, as many organizations have already 'cut to the bone'. Thus, organizations need to manage performance: cost and benefits, as well as increasing market share and customer service through agility.

Executive managers cannot afford to be surprised by their performance at the end of the reporting period. The desk of an executive will become more of a 'control tower' from where the business processes will be run and exceptions flagged instantly. Fine-tuning will happen through extensive what-if scenarios. A prerequisite for this to occur is that the business processes are adequately modeled, describe how staff work, that staff work according to the processes, and targeted performance levels are actually realized.

Our Management by Process Framework model clearly positions process execution and process performance as key aspects of management. The impact and performance of the business processes will determine which projects will be executed. The outcomes of the business processes will be closely aligned with the strategy. Overall governance will be put in place to ensure that the alignment is structural and continuous.

Everything is seen from a customer perspective

Customer-centric end-to-end business process thinking will expand to all activities in the organization. Processes will be more and more viewed from the contribution they make to adding value to the customer and the alignment and realization of the strategic objectives.

This will break from the predominantly held view that business processes should be modeled from inside out. Process improvement will start by obtaining a better understanding of the customer and their demand on the business. A customer-centric view will also ensure that it becomes easier to incorporate business partners as part of an overall customer proposition.

As outlined in this book, we expect that business processes and the process stewards will have more of their KPIs (Key Performance Indicators) relating to customers. To ensure that there is better control in achieving this, it is expected that process stewards will insist upon establishing more lead indicators to better predict outcomes. Customer-centric process models will guide managers in modifying the business processes and what impact this will bring on the customer experience.

Knowledge workers

Gone are the days when only the repetitive and mundane business processes are modeled because they will become largely automated and knowledge-intense processes will be better managed as processes. The main reasons will be that compliance requires controls around these processes and the knowledge-intense processes are often the core of the competitiveness of organizations. They will be better monitored with the knowledge being retained within the organization. A more advanced way of modeling, improving and managing processes will evolve. Key elements will be self-reliance, training, competency and peer reviews.

Process execution and management will require much closer participation of the employees. Both the first wave (Taylorism: with clear separation of 'hands' and 'brains') and the second wave (business process re-engineering by external 'experts') will have disappeared. The knowledge worker will become more directly involved in the performance and improvement of process execution and management. This will bring the often requested, but rarely realized, continuous process improvement and innovation closer to reality.

Processes as the basis for automation

Organizations have increasingly realized that strategy and business processes should be leading and that technology should follow. While technology can be an important enabler of innovation, processes and business needs must accommodate innovation and only then can they provide benefit.

The two main frustrations with business process modeling will be resolved: first, the fact that business requirements for IT development are specified independently from the modeled processes; and secondly, the modeled processes once entered in a workflow tool can no longer be conveniently modified in the process modeling tool. The processes and their documentation will form the basis for BPM automation. The advantages of this approach will ensure that the business requirements are process driven and make use of the available documentation. In addition, the best mechanism to check the validity of the information is that people use the information.

Process modeling has been completed for many years in an isolated set of activities separated from the actual process execution and the automated support. With the increased awareness that processes need to be approached holistically and with the advancement of technology, the automation can be specified, developed and tested from the modeled processes. One of the main challenges to overcome for workflow and business rules engines is to present the process models in a way that is easy to understand and maintain by the process stewards.

This trend is strengthened by the further rise of executable process language so that the automation does not require lengthy development activities and reduces the risk of misunderstanding the process and requirements.

End of the Chief Process Officer (CPO)

This seems to be a contradictory trend, just as more CPOs are being appointed, we are forecasting their decline. The main reason for suggesting this is that we consider the CPO role to be a transitional role ensuring that an organization becomes more process-focused. Once the customer focused end-to-end business process thinking is engrained in the organization, the executive responsible for an end-to-end business process will oversee the continuous improvement process and will most likely report to a Chief Operating Officer. The reason that we still propagate the CPO role in this book is that we see it as an essential intermediary stepping stone in achieving a more customer and process-focused organization.

Our model clearly shows that process execution is positioned as part of management as usual. Many organizations are still at a low-level project management maturity and require temporary additional support to make this paradigm shift. Once the strategy, project and process execution are well aligned and strengthen each other, the Chief Operating Officer (COO) can take over.

Even more internal resources

Organizations have learned from the 1990s that large-scale process re-engineering projects with many external consultants does not work. Organizations have experimented with various models using a higher number of internal resources. It has become obvious to the authors that the most successful model is for there to be a relatively small Centre of Business Innovation within the organization and for the various business units to have subject matters experts that are keen to improve their processes. We do not envisage large numbers of BPM 'Black Belts', but a more wide spread awareness of BPM. Internal capability will be built by increasing the internal skills and knowledge to improve and manage processes, thus creating a true continuous improvement environment.

Once the dimensions as described in this book have been implemented, especially the process leadership and the people dimensions (capability and performance) the overall belief in, and understanding of, the benefits of a process-focused organization will be realized. This will be enhanced further by continuing successful business improvement and management projects.

We foresee an increasing number of employees who want to actively participate in the process execution and management. Furthermore, we have already seen evidence that people with a passion for process and performance are moving to organizations who share that same passion.

Governance as part of process management

Governance and compliance of business processes has increased significantly in the last few years. Many organizations have moved from a situation of scattered governance to a large and elaborate framework of governance. Many of these structures will be difficult to maintain in the long run. We envisage that more and more governance will be included in the main stream of BPM. This will reduce the risks that governance and compliance are isolated 'ivory tower' activities. There will always be the need to have an independent compliance and governance department within an organization, however, the daily validation of governance and compliance will be completed as part of the process management process.

As outlined in our model, governance is at the centre and many would argue the overarching or central dimension. It is crucial that governance is integral to all other dimensions and that all other dimensions are aligned. For example, it is no use focusing governance purely on the execution of projects, while the link to strategy (such as the process of initiating and prioritizing projects) is not governed to the same level.

Case study: Mopping the floor with the tap running

We were asked by an international insurance organization to reduce the number of projects they had from the current level – in excess of 250 projects either in execution or initiation. We were able to reduce the number of projects to about 30 active projects and 50 planned for the next 2 years.

We noticed that a significant number of new projects were initiated. The reason was that a strategic consultancy organization was reviewing the strategy and was actively asking the managers to suggest new projects and ideas. However, they failed to follow the agreed project initiation process and had no effective mechanism in place to rank, prioritize and select projects.

We implemented portfolio governance to ensure that projects were aligned with the strategy and in line with the available resources and budget. As a result the managers involved realized the importance of managing the list of projects.

(Continued)

Case study: Mopping the floor with the tap running (*Continued*)

Message: To ensure that strategy, project execution and process execution are aligned, it is important to have governance in place for these three dimensions and the interfaces between them.

Accreditation

We believe that the process management community is approaching its next level of maturity which will include the accreditation of process analysts, process consultants, process managers (process stewards) and BPM project managers. Several attempts have been made by various firms to develop accredited training courses; however, most of these courses do not provide the sustainability of the training material intellectual property, while others have theoretical content without an all encompassing project methodology model. Some training organizations provide training where their BPM expertise and methodology is only existent in the presentation material and not within supporting documentation. The strength of IT management and project management methodologies, such as ITIL and PRINCE2, is that they have an overall encompassing detailed model with a consistent approach for each module and has been based on extensive expertise.

Our model emphasizes that business processes and process management cannot be completed in isolation and must be linked to execution, strategy and projects. Hence, process skills accreditation should not be limited to only process analysis and metrics, but it needs to include management and strategy people. A last note on accreditation: a track record of real-life success is more important than having a heap of paper knowledge and paper certifications.

Process community

There is a greater need than ever for a process (BPM) community. This is both internally within an organization and externally between organizations. We have seen various attempts at establishing process communities, but the most sustainable are those with no commercial interests and are independent, open for everyone and have the right (and enforced) balance between process users, consultants, vendors and academics – the Dutch BPM Forum is a perfect example of this. It is still growing strong after more than 3 years (www.bpm-forum.org).

Another critical success factor for a true BPM community is the involvement of executive management in these communities. Processes and process analysis is not an objective in itself, but a means of achieving a business objective. Communities that only focus on process analysis or process architecture tend to have many theoretical discussions without providing value to the participants or their organizations.

Embedding in the organization

We believe there will be further embedding of process thinking and management within organizations. The best indicator of this occurring within an organization is the number of times that processes are discussed at the executive level. These discussions could be related to: solving specific business problems (e.g. it could relate to customer satisfaction issues), integration with business partners, and the importance of processes in mergers and acquisitions. In other words, business processes become an important starting point for the discussion of problems and opportunities within an organization.

Our model shows that processes are at the heart of the organization. Business improvements projects will only succeed when embedded within the process execution. The rewards for employees should be related to the contribution they make to the business process outcomes and organizations strategy.

Process leadership

Above we have described ten key trends that we are seeing, or foresee in the near future and are sure there are many more, however, there is a single, overriding critical success factor that is essential for an organization if it is to attain significant business improvement and the management of its business – or things will simply not change. This is *process leadership*.

It has been shown over and over again in the case studies, that without the understanding and drive of a business leader, the battle towards becoming more process-focused (and all the benefits this beings with it) is simply not achievable, or at best, very difficult. Process awareness of the leader, drive and sustained passion is required. While ideally the leader should be the CEO of an organization, it may be possible to have the leader of a substantial business unit drive a process-focused approach within their business area of influence.

While there are definitely some leaders who have "got" process or are "getting" it, we are not seeing the trend as dominate as we would like it to be. It will take time and perhaps needs to start with education in universities on business process improvement and management. These will allow educated professionals to come into businesses and start to push its leaders in the 'right' direction.

As more and more detailed case studies are published and made available to leaders, hopefully they will start to understand the journey and the significant benefits of taking it.

It will take courage, but leaders are not alone on the journey, there are plenty of skilled and passionate coaches and guides available to assist on the journey.

Part III

Appendices

In the appendices we have provided additional information for most of the dimensions that the reader may find of interest. Appendix D is in more detail and comprises our comments on the technology that is applicable to business process management.

Appendix A

Process leadership

Conscious competence learning

Many organizations and individuals fail to understand their low level of process management maturity. Before any effort can be made to improve competency, it is firstly essential for executive management and staff to be aware of the extent of the incompetence. Failing to do this can result in too much emphasis being placed on training people to increase their ability, while they are not open to receive the training, or worse, they may feel insulted because they consider themselves to be fully competent.

The conscious competence learning model of the US Gordon Training International organization is a useful tool in this regard (Johan, 2007).

This model assumes the following stages:

- *Unconscious incompetent* – this can relate to both the under-estimating of the importance of process management and the unawareness of their skills and expertise in this regard.
- *Conscious incompetent* – during this stage training and coaching can be provided.
- *Conscious competent* – the learned insight is being applied consciously with mentoring being a useful activity at this stage.
- *Unconscious competent* – process management becomes embedded in the management as usual.

It is important to recognize that people may eventually fall from being unconscious competent to conscious incompetent. This could occur if they do not keep abreast with advancements in technologies and methods, as well as the circumstances of the organization, such as increased process management maturity level.

Situational leadership

Most process literature either ignores the essence of adequate leadership, or they review it in a superficial manner. It is necessary to take account of the various situations that leaders face during the 'bumpy' ride to achieving success through becoming a process-focused organization. In other words, a successful leader is able to adopt different leadership styles depending on the situation.

The Situational Leadership model of Hersey, Blanchard and Johnson (2005) is useful in this regard as it allows an analysis of the specific needs of a situation, and then the adoption of the most appropriate leadership style. This model has proven to be popular with managers over the years because it is simple to understand, and it works in most environments for most people. The model rests on two fundamental concepts – leadership style and development level.

The model specifies the leadership style on the basis of two dimensions:

- the amount of direction that needs to be provided and
- the amount of support that needs to be provided.

As a consequence the following four quadrants can be specified:

- **S1**: *Directing Leaders –Telling and Directing*: High directive behavior and low supportive behavior.
- **S2**: *Coaching Leaders – Selling and Coaching*: High directive behavior and high supportive behavior.
- **S3**: *Supporting Leaders – Participating and Supporting*: Low directive behavior and high supportive behavior.
- **S4**: *Delegating Leaders – Delegating*: Low directive behavior and low supportive behavior.

The right leadership style depends on the level of the followers. Blanchard and Hersey have categorized the possible development of followers into the following four levels:

- **D1**: *Low Competence, High Commitment* – they generally lack the specific skills required for the job in hand. However, they are eager to learn and willing to take direction.
- **D2**: *Some Competence, Low Commitment* – they may have some relevant skills, but will not be able to do the job without help. The task or the situation may be new to them.
- **D3**: *Moderate to High Competence, Variable Commitment* – they are experienced and capable, but may lack the confidence to go it alone, or the motivation to do it well or quickly.
- **D4**: *High Competence, High Commitment* – they are experienced at the job, and comfortable with their own ability to do it well. They may even be more skilled than the leader.

It is important to notice that in the context of business process innovation and management the relevant development levels should be taken into account. Thus, the development level does not relate to the generic ability of the followers, rather to the specific process and innovation ability and skills of the followers.

Thus, the leader has to adjust to the situation and the followers. Not all leaders are able to perform all four leadership styles. Furthermore, some leaders actually have a strong preference (and ability) to be successful in a particular situation.

From a corporate level it is important to plan the deployment of managers accordingly. Directive leaders should be succeeded by leaders more aligned with coaching and supporting.

Appendix B

Process governance

Roles and responsibilities

Process steward

In summary, the types of skills that are considered to be essential to be a successful process steward are:

- being able to envision and understand the big picture, while having a detailed understanding and perspective of the end-to-end business processes you are responsible for managing.
- being continuously customer-aware and understanding, while clearly understanding the customer needs.
- being intimately aware of the organization's culture, and ability and capacity for change.
- being a logical thinker and having an ability to analyse data and situations.
- communication – the need to sell the benefits of a business process–focused approach to all stakeholders.
- negotiation – to ensure that all stakeholders, especially the functionally based stakeholders, have their needs and issues addressed or where compromises are required, that there is understanding and acceptance of the compromises.
- results orientated/focused – at the 'end of the day' the organization needs to achieve its strategic business objectives to the satisfaction of all stakeholders or there will be no organization in the long term.

Process executive

The skills required to be a successful process executive include the following, but certainly do not limit it to these skills only:

- Leadership (commitment, time, 'walk the talk'). Unless the executive is passionate about taking on the role of process executive, then you are not the 'right person for the role'. It is not just *part* of your day-to-day role; it is a *significant part* of the day-to-day role or indeed, *all* of your role. You need to participate in the Strategic Process Council to

- be knowledgeable of all the business processes in the organization and
- understand how the main business processes intertwine.

- Maintain a strategic focus and understanding – how the main business processes specifically link and add value to the organization's strategic business objectives.
- Be customer-aware and understanding – the customer comes first and last. This also means understanding and balancing business objectives with customer needs.
- Communication – the need to sell a business process–focused approach to all stakeholders.
- Negotiation – to ensure that all stakeholders, especially the functionally based stakeholders, have their needs and issues addressed or where compromises are made, that there is understanding and acceptance of them.
- Results orientated/focused – at the 'end of the day' the organization needs to achieve its strategy and business objectives to the satisfaction of all stakeholders or there will be no organization in the long term.
- Work with partners to optimize the business processes and the ability to work with vendors.

Business cases

As discussed in Chapter 5, a business case is a critical aspect of the allocation of organizational funding and project prioritization. We have outlined the various aspects that relate to a business case below.

Why do you need a business case in the first place?

Different participants or stakeholders in a project will have different perspectives on why you need a business case. A project manager's perspective often includes the following:

- The need to get the project approved in the first place, so that the project can commence.
- It will assist in gaining access to the various resources and funding required to complete the project.
- It will ensure that the project has the initial 'buy-in' and support from the process steward and sponsor/executive.

- It will provide guidance for the execution of the project throughout the projects lifecycle.
- It should ensure that the project is well planned, scoped and communicated to all stakeholders.

From the business' perspective, the business case will:

- Determine if the project can be justified commercially, that is, is it worthwhile executing in the first place.
- Determine if the project makes sense from a business perspective and is in alignment with the progression of the organization's strategy and business objectives.
- Measure how much value the execution of the project will add to the organization.
- Help the executives to also look at this project in relation to other projects submitted for approval to ensure that it is the best use of an organization's available resources (time, money and people) and capacity.

A business case will also show if the project has been well thought through and planned, and that the project manager and process project steward/executive/sponsor know what they are doing. The business case also needs to answer the questions: Does the business have confidence in the project? Is the project governance appropriately established? Have the project and business risks been defined and are they acceptable? Have effective risk mitigation strategies been put in place?

Unfortunately even organizations that use business cases for the approval process rarely use them to *manage* the project and to evaluate the success of the project.

Once approved, the business case fulfils the following purpose during the three stages of the project.

At the start

As stated above, the first task of the business case is to ensure that the project can be commercially justified and to gain approval.

During the project

The project scope, contained within the business case, will define the boundaries and deliverables of the project. These must be in the forefront of the project manager and project steward/executive/sponsors mind at all times. Whenever requests for changes or additional functionality are suggested, they need to be referenced against the scope to determine whether or not they fall within the project scope. Obviously, changes and additions that fall outside scope need to be approved via the project change management procedures. The project manager must continually monitor the likelihood of realizing the benefits in the business case. If it is clear that the benefits have substantially reduced or disappeared, then the project must be stopped and referred to the process steward and/or process executive (and project steering committee) and perhaps the Strategic Process Council for a decision. Any project scope change request raised should be attached to the Business Case. These benefits must also be continually referenced throughout the project to ensure that there is no slippage, and if slippage is expected it needs to be documented and communicated to stakeholders in the same way as a change of project scope. (Refer to Jeston and Nelis, 2008, Chapter 21, for details on an effective way to document benefit changes and control benefits realization management.)

After implementation

Most project managers consider a project complete once it has been implemented – this is not the case. Projects are initiated with specific defined benefits within the business case, and a project is only complete once these benefits have been realized within the time frames specified in the business case. The business case is the basis to measure the success of the project.

Business case summary

The writing of the business case is one of the most important activities in any project. If written well and then used correctly throughout the project, it has the ability to both guide the project and make the project's completion easier. There are various types, levels of complexity and size of business cases. Depending on these aspects, the business case could take a couple of weeks and a few thousand dollars to write, through to six to twelve months and up to a million dollars or more.

As a final check, the project manager and sponsor must ensure that the business case has answered the following questions:

- How well does the proposed project meet the strategic objectives (imperatives) of the organization?
- Which alternatives have been considered and what were the outcomes of the various options?
- Are the project costs robust and accurate?
- Are the business benefits outlined robust and accurate?
- Can this project achieve the targets shown in the business case within the current situation and risks?
- Who is accountable for the accuracy of the costs and business benefits?
- Is the project worth the money, from the business perspective?

Getting it 'right' is critical and the project manager must spend the time at the commencement of the project or the project will struggle from the start.

There are three critical success factors for any project:

- Manage the business case
- Manage the business case
- Manage the business case.

This must occur from start to the completion of the project, with a need to ensure that the outcomes of the project are continually in line with the expectations of the project sponsors, stakeholders and business case.

The last question to ask: 'Is a business case useful?'

The answer to this question is covered by the following four points:

- A strong business case *equals* proper funding.
- Proper funding *equals* reduced risks.
- Reduced risk *equals* more successful projects.
- More successful projects *equals* better business value.

A strong business case framework is a critical part of any governance structure and project approval process; so make it part of the process governance framework for an organization.

Different types of business cases

There are various types of business cases and we will split them into simple, traditional and unique categories. The type of business case selected will largely depend on the type of project you are undertaking. In their simplest form business cases are either small, large, simple or complex. Refer to Figure B.1 which shows the mix that can occur. Some small projects can be incredibly complex in their execution, while some large projects can be simple. Although, as a generalization, the longer the project the higher the risks and possible complexities. We will discuss this in more detail later.

Table B.1 provides a matrix of both the type of project and the likely characteristics that the business case will comprise.

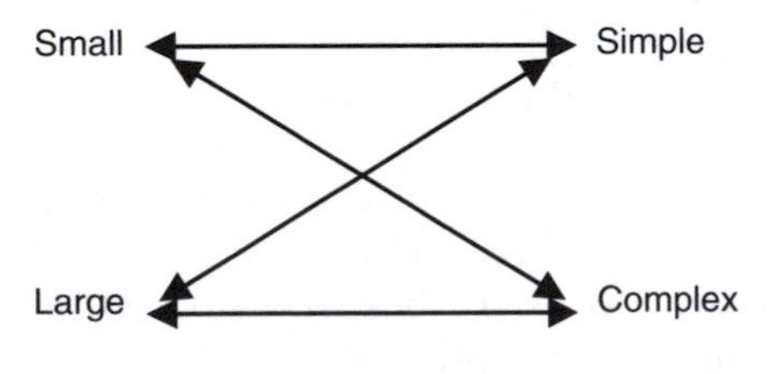

Figure B.1 Types of business cases.

Project selection

Projects are initiated (triggered) in one of three ways. They are either *strategically driven, business issue–driven* or *a process-driven* project. Figure B.2 shows these three triggers as well as the likely courses for the project (Jeston and Nelis, 2008).

If a project is *strategically driven,* it will have been initiated as part of the organization's strategy. The organization will have created its strategy for the next period (usually 12 months or more) which will require multiple projects to be commenced to implement the strategic objectives. Projects of this nature will usually be more easily justified, from a business case perspective, than other types of projects. The business case will need to clearly show the link to the strategy execution.

Table B.1
Project type and characteristics of business case

Type of project	Characteristics of business case
Infrastructure	Technical risk: Vendor performance to deliver when needed. Integration of components. Easy to monitor and measure deliverables.
Package implementation	Cost and scope of risk. Vendor's ability to deliver, on time and on budget. Functionally fit with the business requirements (only 'really' find out as the system starts to be used). Scope and change management. Enhancements requested by the business, particularly around the number of enhancements and the businesses ability to specify them clearly. Impact on the business of simultaneously running the business while implementing the package (need to specify in the business case how this will be achieved and allow for the costs involved). Processes that the package imposes, are they suitable for the business?
Bespoke system development	Ability of the business to specify their requirements. Risks associated with estimating the cost and time associated with the development of the new bespoke system. The business will change as the system is specified, developed, tested and implemented. The business case needs to plan for this and inform the stakeholders how this will be catered for.
Legacy system shut down Legacy system consolidation	Unknown users (we have been involved in this type of project and there was no way of determining how many users there were and in what part of the business they resided). How will the business cope if its users suddenly loose the system? The project manager and project sponsor need to clearly understand that the project is not completed, nor the benefits realized until the legacy systems have been deleted from the hardware and contracts terminated with the suppliers.

Whereas, *business issue–driven* and *business process–driven* projects will need to compete for the limited funds within an organization and therefore need to clearly quantify the costs and benefits associated with executing the project. The project manager and sponsor will usually need to persuade the project approvers both verbally and with a very strong business case. However, a *business issue–driven* project, especially if a compliance or legislative requirement, will usually be more easily justified.

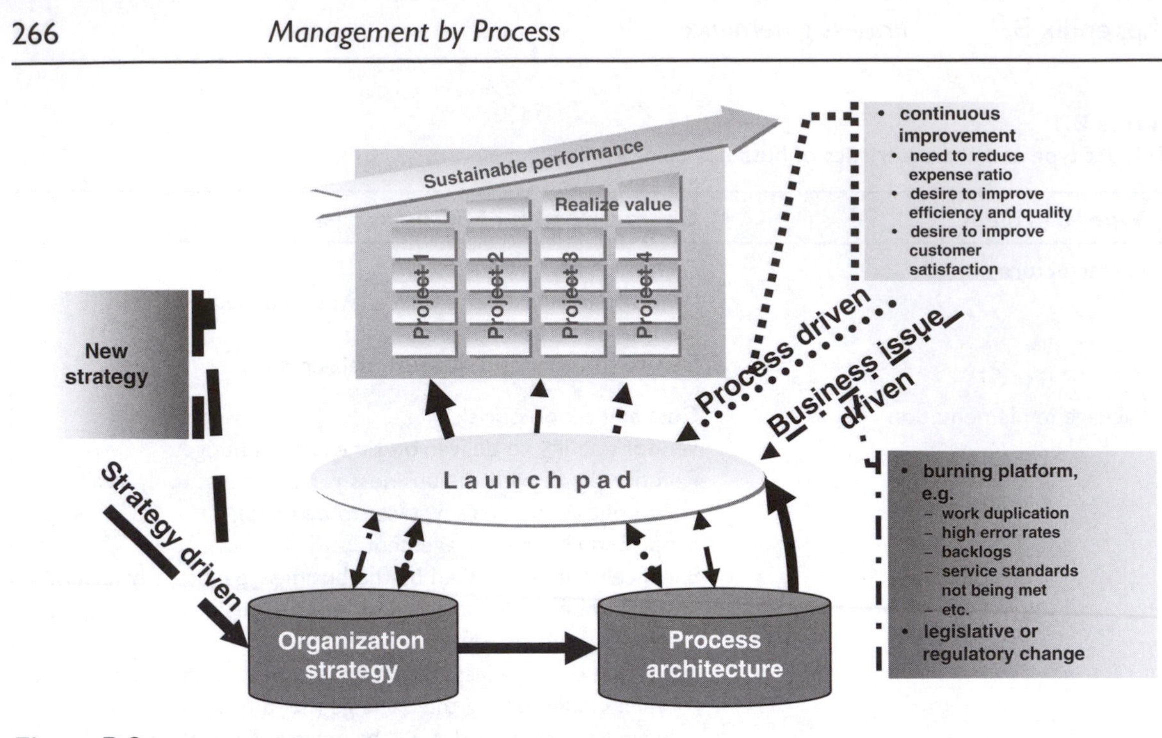

Figure B.2
Triggers for project approaches (Jeston and Nelis, 2008).

A *business process-driven* initiative is usually commenced with a small business case to have a 'look and see', small discovery project, to determine if there are opportunities for a full-scale project. This will usually take a 'gated' approach, as discussed in Chapter 9, where small amounts of project costs are approved. Once the review is completed it will provide sufficient information to justify the next stage of the project. Thus, the business case is completed and approved over the life of the project.

Project scenarios

Apart from the project selection mentioned above, there may also be different types of project scenarios as shown in Figure B.3 (Jeston and Nelis, 2008).

We will discuss these four different scenarios and specifically relate them to both a process-focused organization and the business case needs.

The scenarios are a function of the involvement of the business manager and the impact of the project on the organization.

Involvement of the business manager relates to how comfortable the business manager is with the particular type of project being put forward, for example, a business process improvement/management (BPI/BPM) project. Has the manager had experience with this type of project previously? Does he or she believe BPI/BPM can deliver significant benefits to the organization?

The vertical axis relates to the impact the project will have on the organization. Is the project a small discrete project with little impact on the overall organization or will it impact significant parts of the organization? This could also be related to the maturity of the organization with regard to conducting process improvement projects. Has the organization significant experience

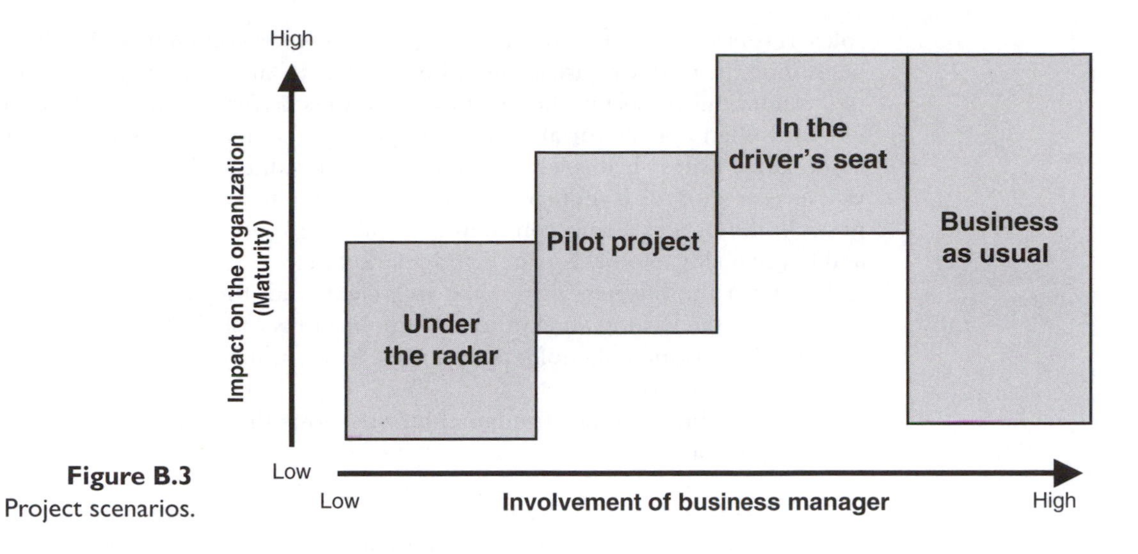

Figure B.3
Project scenarios.

with conducting successful business improvement projects? Do they need to rely on external resources (project managers) to manage their projects? Are their project sponsors trained in their role and responsibilities?

Each of the scenarios is now explained (Jeston and Nelis, 2008).

Under the radar. This occurs in the least mature organization and is where there is a partially informed business manager who is not yet committed and is not paying much (or any) attention to BPM within the organization. This scenario could be a project under the guise of process improvement and BPM may not be mentioned at all.

Pilot project. This is where there is a fully informed business manager who is yet to be totally convinced of the benefits of BPM and is willing to try it out on a small scale to start with before making a full commitment.

In the driver's seat. This is the next highest level of organization maturity of a process-focused organization, and is where there is a fully informed business manager who is totally committed to the implementation of BPM within the organization or business unit.

Business as usual. This will be selected by the most BPM and project mature organizations. The organization and business managers will be totally committed to a process-focused organization, and BPM projects are simply business-as-usual activities or projects.

The impact on the business case for each of these scenarios will range from having to do no convincing of the merits of BPM in the case of *business as usual* through to a great deal of convincing for the least mature scenario in *under the radar.* The business case will still need to show the cost/benefits in its own right, but the level of scepticism and confidence in the project will vary according to the understanding, maturity and impact the project will have on both the business manager and the organization.

Business drivers

The project manager is often the person who writes the business case, supported by the project sponsor from the business – we would rather see these roles reversed. Another aspect that needs to have complete clarity before beginning the business case is the business drivers that are causing the business to consider this project in the first place. Business drivers can be many and varied, and often appear (or are) conflicting. Is the project necessary to address compliance issues? Is it solely motivated by cost reduction? Are increased service levels required? Is there a need to make a step increase to obtain new business? Is the project being considered to provide the business with a level of agility (an ability to quickly react to the market place)?

Whatever the business drivers, be very clear what they are as they will not only impact the writing and content of the business case, but also the 'selling' of it to the various stakeholders and then how the project is evaluated with regard to its success.

There are however two fundamental questions that must be answered in any business case:

- How is the project going to add value to the organization?
- How does the project align with, and contribute towards, the organization's strategy?

Unless these two questions can be answered clearly and concisely, then the project should not be approved.

The type of project will have a significant impact on the way you write and present your business case. So spend the necessary time up front gaining agreement amongst the various stakeholders on the type of project that is being proposed before commencing to write the business case.

Complexity versus time versus risk

There has been a great deal written on risk in many books and we do not propose to write about it here, other than to show a couple of impacts upon risk that are worth covering.

There is an interesting relationship between the three components of complexity, time and risk and Figure B.4 shows it clearly.

As a generalization, there is a very strong relationship between the length of a complex project and associated risks, and risk usually increases in an expediential manner. If a project is complex and is likely to span over a lengthy time frame, then perhaps it should be broken down into a series of less complex and shorter projects. Milestones and progress are more easily managed, as are the risks. Therefore, perhaps the business case should be for a programme rather than a project.

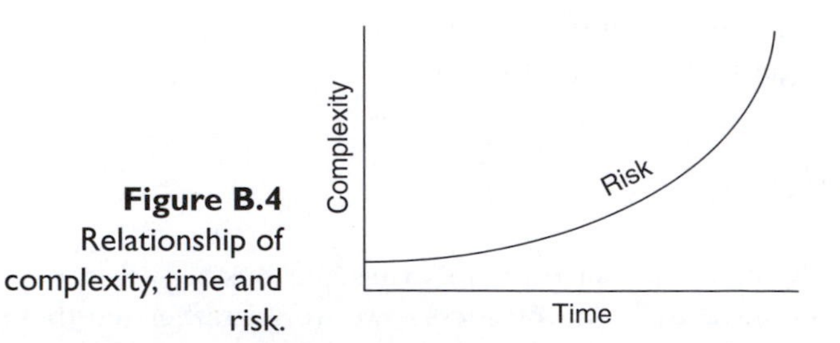

Figure B.4 Relationship of complexity, time and risk.

The business case must analyse the risks associated with the project and outline the mitigation strategies. Most risk analysis registers will cover at least: a brief description of the risk; the likelihood of it occurring; the impact upon the project; the owner or person responsible for addressing the risk; and the proposed risk mitigation steps or strategies. While this is necessary, we often find that busy executives just glance over the risk register or ask the project manager is there anything they should be concerned about. A technique that we have used is to show the risks visually in the business case and subsequent steering committee meetings as shown in Figure B.5.

Each number in the figure represents the risk number from the risk register and it is then placed in the appropriate quadrant.

We have found that by showing the project risks in this format provides a picture that is easily understood by busy people. Obviously it is the upper right-hand quadrant that should be of interest to the stakeholders. Where we have considered a risk to be so important that we needed to get significant attention placed upon it, we have been known to place the risk number outside of the upper right-hand quadrant and then highlight it in red.

How do you 'sell' your business case?

Well, you have done the hard yards. You have determined from the myriad types of projects what type of business case to write. You have determined the orientation and appropriateness of the business case for the organization and now it needs to be 'sold' to the stakeholders. When should you start stakeholder management and how is this achieved?

We would suggest that you should have started quite some time ago, when the project was first conceived. Unless this is the case, you will not know if you have answered all the objections or barriers likely to put up by the stakeholders.

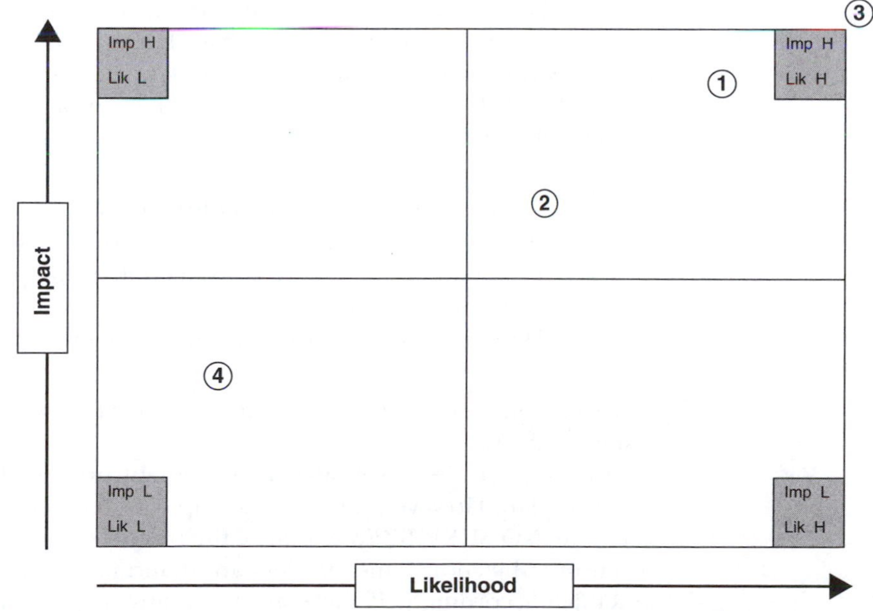

Figure B.5
Project risk analysis.

Case study: No one wanted the business case to succeed

We were requested to write a business case for a large government organization. The project sponsor, in our opinion, only wanted to write the business case to protect himself from future criticism because he 'knew' that it could not be justified. He openly stated that it would not work because they had tried it four years previously and the staff rejected it and the implementation had to be withdrawn. The chief information officer knew that his staff was extremely poor at delivering projects and hence would have preferred the business case not to succeed. It was obvious to the business that the implementation of the proposed workflow and imaging system was the only way they could survive. Their volumes were about to increase by 600–700% (a change in legislation) and the projected influx of paper could not physically be stored in the building, let alone processed. We received no assistance in the writing of the business case from the organization, other than the part-time assistance of a business analyst.

The first thing we did was to interview the project sponsor, followed by a group workshop with the CIO, IT staff and the business personnel. We asked them all only one question?

'What does this business case have to answer for you to approve the project?'

This approach elicited all their objections. They told us all the things that would have to be overcome for them to give the project their support.

Once complete, the project sponsor read the business case and stated: 'well, this is a bit of a no-brainer, isn't it?' We had not only addressed all objections, but shown them the way forward. We then project-managed the implementation of the project and it was openly acknowledged as one of the organizations best project successes.

Message: If you do not clearly understand stakeholder expectations and objections up front, you cannot address these in the business case and have a high probability of gaining approval.

As indicated above, stakeholder management is the way you 'sell' your business case. Managing stakeholders takes an extraordinary amount of time and focused effort. It is not something that should be completed by the 'seat of your pants' or by intuition alone at the time you are meeting with a stakeholder.

Stakeholder management is acknowledged as a managed process in its own right. It is something that needs to be worked through, discussed, planned and then implemented, much like a project. The usual high level steps involved are:

1. Establish an internal stakeholder team
2. Identify stakeholders and their relationship with the project
3. Profile the role that the key stakeholder will play in the project
4. Map the stakeholders
5. Determine a strategy to engage and manage them.

Jeston and Nelis (2008) in Chapter 24 show this stakeholder management system in detail.

It is only via the stakeholder management process that you will 'sell' the business plan. However, there are a couple of *must do's*. First, the golden rule is to have NO SURPRIZES. Socialize the business case as it is being developed to ensure objections, and barriers are determined, and the business case is amended accordingly. If there are to be any surprises or dramatic changes, ensure they are addressed as early as possible, before it is made public and certainly not in the last minute before the business case is submitted for approval. The project manager and sponsor must know before they go to the approval meeting that they have everyone's positive vote.

A project manager/sponsor should also be aware of the political environment within the organization, but must not become part of it.

What do you do when you cannot build a business case?

Assuming that the difficulty is not to do with stakeholder management and all stakeholders have been involved in the business case process, the difficulty could be that there is either:

- Significant costs involved in the completion of the project and no immediate and measurable benefits for the business; or
- Non-financial benefits that cannot be justified against the project costs.

Significant costs and no immediate and measurable business benefits

One way to overcome this is to create a basket or programme of related projects which include your project. This can be achieved by preparing a series of business cases and rolling them up into an overall programme that delivers positive value to the business.

Examples where this could be the case include:

- a desktop replacement project, where the project is essential to the long-term viability of the business, but has no direct positive value in its own right
- showing how the desktop replacement project will enable the delivery of other value adding projects, such as the delivery of a single customer view.

Non-financial benefits

Projects of this nature have no financial benefits in their own right, but may be justified by the avoidance of downside risks and the consequences of doing nothing. Examples include:

- The hardware and software are obsolete and are no longer supported by the vendors. There will be a need to move onto other hardware and/or system.
- The technology does not enable disaster recovery or business continuity planning activities. This will result in there being a high risk of permanent failure and loss of data and documents.

- A legacy imaging system that contains customer critical data and documents, such as bank authorities, loan and investment documents.

At the end of the day, perhaps the project cannot be justified at all. The project manager should not feel like he or she has failed. One of the main purposes of the business case is to determine if the project is financially justified and will add value to the business or organizational strategy. An answer of no can be the right answer for the business.

Appendix C

People capability

Roles and responsibilities of CBI

The roles and responsibilities of the Centre of Business (process) Innovation (CBI) are outlined below and supplement Chapter 8. The structure could look like that shown in Figure C.1.

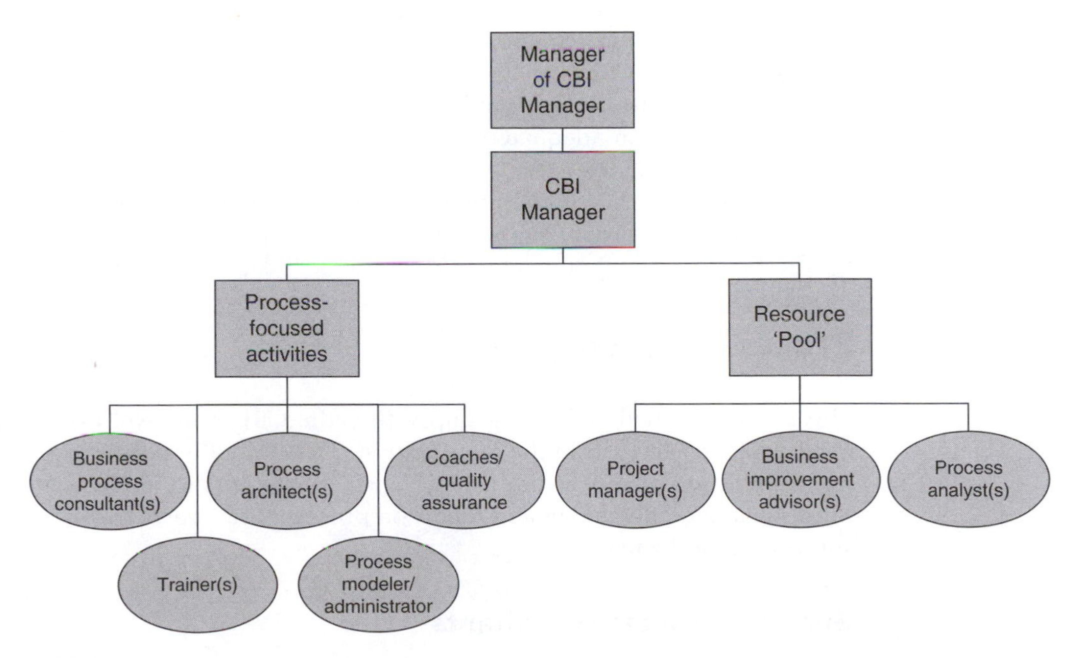

Figure C.1
Sample centre of business innovation structure.

Manager of the CBI Manager

This role has the following responsibilities regarding processes:

- Ensure that the strategic alignment is in place so that the deliverables and work of the CBI group is aligned with the objectives and expectations of the business.
- Ensure that the CBI group continues to contribute to the achievement of the objectives of the business particularly in relationship governance.
- Assist the CBI Manager in the execution of his or her role.

CBI Manager

CBI Manager has the following responsibilities:

- Lead and drive continuous process improvement design and approach across the business.
- Assist business management in the creation and maintenance of a high performance management environment.
- Development of the process architecture, ensuring that the business processes are continuously aligned to the organizational strategy (objectives). Furthermore, ensure that IT is aligned to these business processes.
- Stakeholder management and expectation management for process management and project activities.
- Quality management and the satisfactory involvement of the staff.
- Lead the CBI team, especially in gaining action or agreement from other parts of the organization.
- Ensure that adequate resources and facilities, of all kinds, are available to the team.
- Assist in the development of business cases and business impact analysis.

Process-focused group

The process-focused activities group within the CBI group will not become involved in projects within the organization. They will offer consulting, coaching, mentoring, quality assurance and advice for others to implement. They will also monitor, document and report on process governance, lessons learned and project performance.

Business process consultants

This role can also double as the account manager for a part of the business. The role works with the business to identify the opportunities for process improvement and process management and coordinates this with the CBI. The business consultant should be the first person to discuss process improvement opportunities with the business, and the role that the CBI will play. They can work with the operational business managers to coach them in how to manage their business processes and the type of targets and measurement to implement.

Business process consultants will be responsible for the implementation and adherence to the agreed organizational process governance method.

Process architect(s)

This role ensures that the process architecture is formulated, updated and is being used. The process architect(s) will also be closely involved with the formulation of the enterprise architecture.

Coaches/quality assurance

This role provides coaching to the various members of a project team, particularly the process analysts, business analysts, process modellers and business improvement advisors. However, they will not 'do' the work – only offer advice, coaching and direction to the project team. They will also complete quality assurance activities and reporting on the various projects in progress.

Trainer(s)

Deliver process training to CBI team members and the business. This role may be fulfilled by one of the business process consultants, or an external organization.

Process Modeler/Administrator

This role reviews the process models produced by the business and projects. They ensure that the modeling standards are being observed and that the models are of a standard that can be moved into the central repository as completed. This role will make recommendations for improvements or changes to the process modeling standards of the organization. They will also maintain and administer the central process repository.

Resource pool

The *resource pool* group within the CBI will simply be a pool of resources for projects and the business to utilize. The CBI manager will work with the business to allocate these resources in an appropriate manner.

CBI project managers

These staff should only be used on business projects that have a significant process component. While it could be argued that all business projects *should* have a business process component, sometimes the need for specialist process knowledge is not great and some projects, for example, infrastructure implementation, may have no process involvement. The responsibilities of a CBI project manager should include:

- Day-to-day management and execution of their project.
- Ensuring that all people change management, human resources, business process and training issues are addressed and implemented.
- Ensuring that the project is in alignment with the organization's strategy and objectives, and notifying executive management once this is not the case.
- Management of all activities associated with the project to deliver the requirements of stakeholders in the planned time frame, budget, quality and deliver the business benefits outlined in the projects business case.
- Preparation and tracking of the budget.
- Gaining commitment from all stakeholders for the project.
- Coordinating and gaining agreement for project plans.
- Reporting project progress to all stakeholders on the agreed time frames.
- On-going communication to the organization, IT and management.
- Identifying, managing and elevating, if appropriate, potential or existing issues and risks that may, if left unchecked, impact the project.
- Seeking the assistance of the CBI Manager and a business process consultant where required.

Business improvement advisors

Within the context of the business, this role has responsibility for:

- Identifying detailed areas for improvement within the business.
- Discussing and facilitating the incubation of ideas and changes.
- Specifying and communicating the benefits of the ideas and changes as part of the business case, or why the benefits do not yet exist for certain ideas.
- Specifying the proposed new redesigned processes.
- Applying the process architecture.
- Developing business cases and business impact analysis.
- Providing assistance to project managers, process analysts and the business as required.

Process analysts

Responsibilities for process analysts include the following:

- Model and document the current and proposed new redesigned processes.
- Obtain metrics and calculate the costs for the business processes.
- Complete people capability matrix analysis and recommendations.
- Perform process analysis and business impact analysis.
- Write and specify appropriate documentation as required by the project manager.

Appendix D

Technology

Introduction

While we consider the use of technology to be one of the seven dimensions (Figure D.1) in the drive towards a process-focused organization, it is and can never be the driving force. However, much of the process management literature focuses strongly on Business Process Management Systems (BPMS).

We first want to position automation within the BPM context. The streams described in Figure D.2 have, in general, been described separately.

- *Process-thinking*
  The initial Scientific Management (Taylor, 1911) focused on time and motion studies to find the 'one best method' of performing a task.

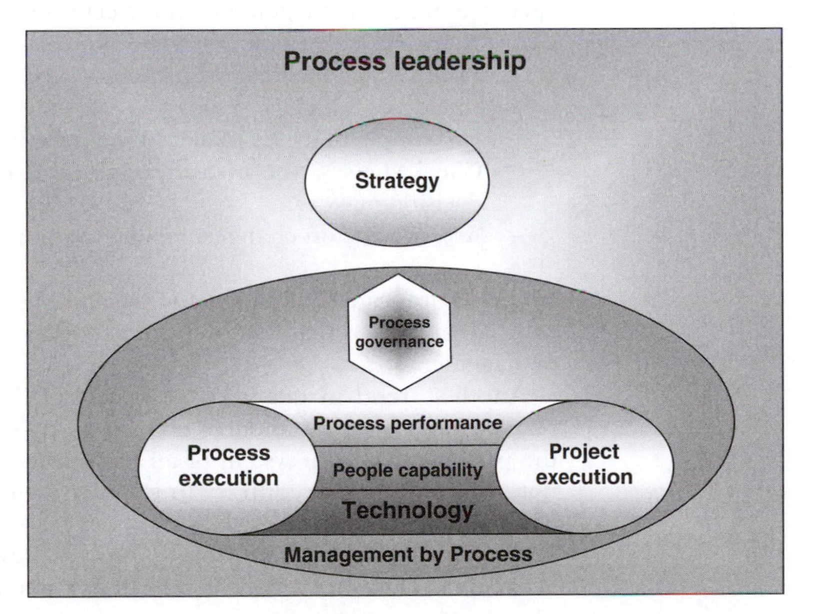

Figure D.1 Management by Process framework: Technology.

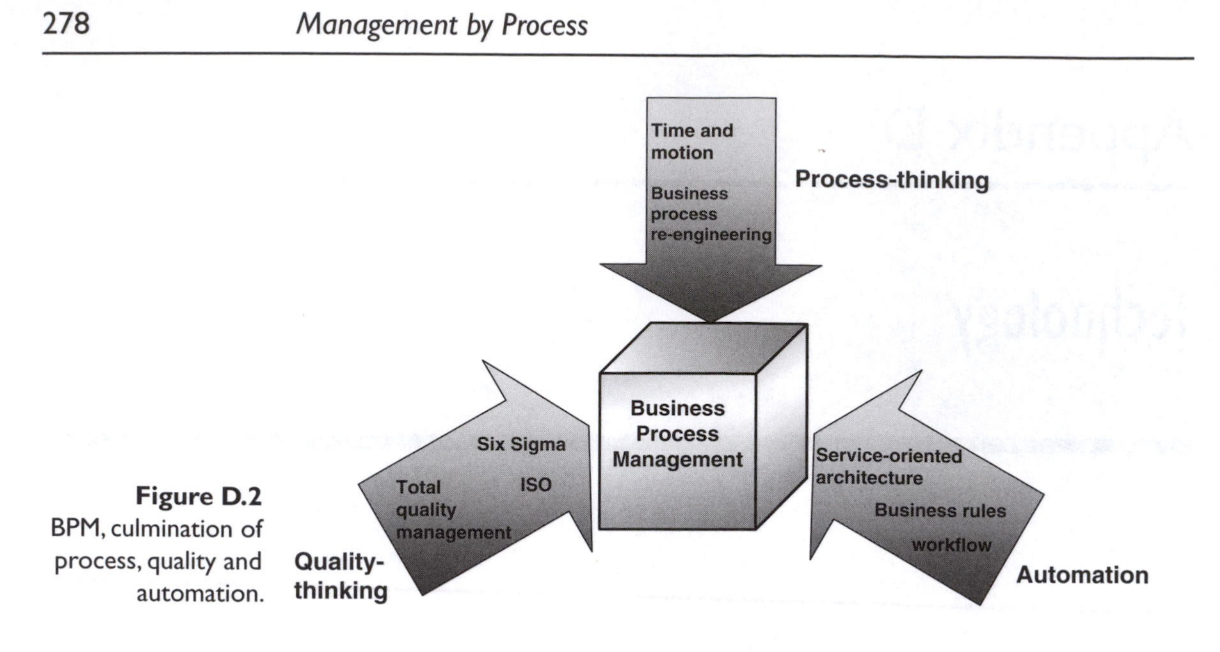

Figure D.2
BPM, culmination of process, quality and automation.

There was a clear division of labor between management and employees – the management had to 'think' and the employee had to 'work'. This was geared towards assembly line work with a high degree of specialization and limited empowerment of the employee. The first assembly line of Henry Ford was based on these principles.

In organizations that were information and service based, this division of labor led to significant overheads, error rates and extended lapse times. Hammer and Champy (1990) outlined how to re-engineer the corporation when they advocated more empowerment of employees and reducing the number of hand-offs as related tasks would be completed by the same individual.

- *Quality-thinking*
  Deming initiated the Plan-Do-Check-Act approach that positioned quality as part of management and execution:

  - Plan – design or revise business process components to improve results
  - Do – implement the plan and measure its performance
  - Check – assess the measurements and report the results to decision-makers
  - Act – decide on changes needed to improve the process.

EFQM (European Foundation for Quality Management) has developed an approach that combines quality management and process management.

Another trend in quality was formal certification. The Quality Management System standards created by ISO are meant to certify the processes and the system of an organization. It is important to note that it does not, contrary to popular believe, certify the product or service itself.

Six Sigma relates to improving processes by reducing defects. Six Sigma relates to a failure rate of 3.4 per million, or 99.9997 per cent perfection! It involves a systematic and analytical process of identifying, anticipating and solving problems. It has numerous powerful statistical tools to measure the performance of processes and to identify areas for improvement.

- *Automation*
  The Office automation development can be categorized by several key developments. Each of these has developed over a long period and has had multiple breakthroughs. Given the tremendous development in information and technology we cannot provide a complete overview of all aspects, so we intend to highlight the key developments and how they have impacted the current BPMS.

  - *Digitalization*: This related to the automation of individual tasks, such as typing. Initially these tasks were isolated and related to the entry of individual pieces of information, such as letters and documents. This allowed the individual tasks to be completed better and faster – it also allowed for the duplication of the information more easily. At a later stage digitalization ensured that physical documents could be scanned and processed automatically.
  - *Volume*: The next development was the processing of data in large back-office facilities, the famous data-processing centres. They produced a large amount of simple and standardized transactions in batches and started with the punch cards in the 1940s – check-processing was a typical example. This had a tremendous impact on the type and number of employees involved in this, often mundane work.
  - *Real-time*: The next development relates to the introduction of real-time data processing, both for back-office processing and for an increasing number of front-office applications. The users had access via terminals to the mainframe application. It enabled them to have access to real-time status of data, such as the balance of a bank account. Desktop applications emerged and allowed relatively smaller applications – sometimes they were even developed without the IT department. Basic workflow was later introduced. It was mainly focused on standard processing of simple procedures.
  - *Individual applications*: With the introduction of the Personal Computer in the 1970s, the users had a wider choice – either by choosing their own standard application or by developing their own applications – especially Microsoft Excel and Access were popular tools, and still are today. The threshold to develop stand-alone applications was reduced significantly and quality assurance (including testing and documentation) was mostly neglected. This was the start of an increasing number of applications and various levels of quality and reliability. Many organizations have reached a situation whereby they have more applications than people! (We know of one organization that has 5,000 consulting staff and 6,000 applications!)

- *Flexibility*: The next development was the further additional 'intelligence' and 'agility' in systems with the introduction of business rules, simulation and business activity monitoring (BAM). This made the applications more understandable by, and supportive to, the business. The business was able to use these modules to become more flexible and meet the ever-increasing changes and requirements of customers. Furthermore, the systems started containing less-rigid information, such as knowledge management.
- *Communication and collaboration*: The next development was the ability to integrate internally and externally through the intranet, mobile phones, web services and service-oriented architecture (SOA). It allowed the various business applications in the organization to truly collaborate and also extended to staff in the field with real-time access to information, and the ability to enter and modify data from their PDA's (e.g. on-line sales forms that feed directly from a remote device to the main business application(s) in real-time or near real-time). This also extended to business applications with partners and vendors (e.g. an instant real-time credit-check by a clearing agency).
- *Control*: The current developments are providing control over the vast technological possibilities and business applications. For a long time, technology has been pushed by software companies and organizations are now not jumping immediately to the latest new releases and updates. The year 2000 challenge has been a wake-up call for many organizations. Furthermore, organizations have realized more and more that technology itself can rarely be a sustainable competitive business advantage – it is the application or execution of the technology that provides a competitive advantage – but even this is not enough, unless the organization has the discipline and control to adhere to the technology execution process.

Control relates to issues such as:

- the number of business applications and how well they integrate.
- the sheer volume of data and how it can be properly managed and streamlined.
- stability of applications and developments.
- cost and resources for maintenance, and new development (invariably in most organization new development and bug-fixing reduces the ability to adequately run and maintain applications in the long run).
- Security and privacy issues.

BPM as described in the third wave (Fingar and Smith, 2004) combined these three aspects into one. We have found that the concepts are clear and logical, but organizations take time to truly embed this thinking and behavior within the organization. This model can be coupled with our Management by Process framework model to ensure that all processes are aligned with the strategy.

Road map

In Jeston and Nelis (2008), we have argued that it is important, indeed necessary, to improve your business processes before automating them. We have also stated that, in our experience, up to 70% business process improvement may be gained, before the use of any technology. We still hold this view.

However, are we a fan of technology from a BMPS perspective? Absolutely. Automation will contribute significantly to the efficiency and effectiveness of an organization's key business processes, provided all the automation components are closely aligned and provide a holistic service and support to the organization.

Technology is a support mechanism for business processes and allows the integration of the key business processes with customers, partners and vendors. The new BPMS technology is agile (if implemented correctly) so that it can support both internal and external necessary changes.

Management must realize that they have full control over the value and operation of the technology they implement.

Business process management systems (BPMS)

In this section we will introduce and describe each of the components of a BPMS – refer to Figure D.3. In each section we will provide:

- A brief description of the component
- The benefits of the component to an organization
- Any issues to consider when implementing in the business as usual operations. This will not include any project considerations as this is out of scope for this book.

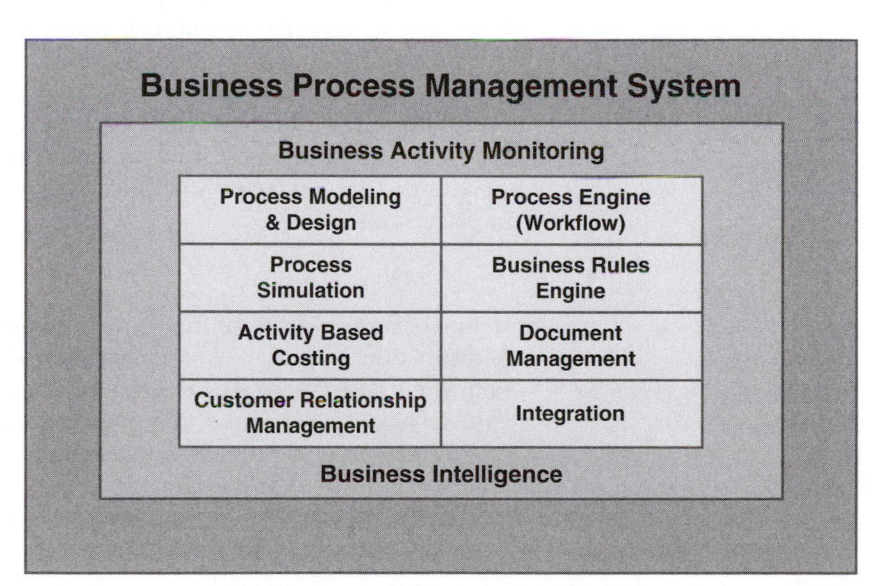

Figure D.3
BPMS components.

Process modeling and design

- *Description*
  This is where an organization models (maps) its business processes and sub-processes. This does not necessarily require a specific BPMS technology tool as it could be done with pencil and paper or can be completed in several of the Microsoft products (e.g. PowerPoint or Visio); however, it will take longer and be far less flexible and maintainable. A technology-based modelling tool will be significantly more efficient. The tools available range from unsophisticated tools that record a business process in a simple format, with no links to other processes; to tools that are extremely sophisticated, linking processes, sub-processes, an overview of an organization, high level value chains, the re-use of sub-processes, on server-based central repository technology.
  A checklist relating to the selection of such a tool is provided in Jeston and Nelis (2008) Appendix L.
- *Main benefits*
  - Provides guidance to the employees on how the business processes should be performed, especially for new employees.
  - Publishing the business processes on-line allows all related information (e.g. current templates, web pages, telephone directories and guidelines) to be included which results in more staff using the process models. This ensures that everyone has access to the most up-to-date information and documents.
  - Provides managers with insight into the business process(es), so that they can manage them and the related resources in a better way. The business process models will also help with the ongoing management of the business process interfaces.
  - Process documentation can be used as a baseline for improvement. Up-to-date process documentation is also beneficial for automation, as business analysts and application developers have a much better insight into what the business process is and how automation can assist.
  - Documentation of the business processes allows the organization to comply with internal and external regulations. The more the information is being used by staff, the more staff is familiar with the processes and the more accurate and up-to-date this information is.
- *Key considerations*
  - Define clear ownership of the process modeling application tool. A distinction should be made between the ownership of the individual business processes contained in the application tool (which should be assigned to a process steward) and the ownership of the process modeling application tool itself, with all its related conventions and architecture.

Case study: Ownership of process models

A newly appointed manager of a Centre for Business Improvement wanted to make a *big* improvement in the organization. He started to enthusiastically model all the key processes. He came up with some seemingly interesting improvements. When asked about who should sign off on these new improvements he stated that as manager of the process modeling tool he was the owner of all the business processes! As a consequence, he alienated the key stakeholders – the business owners of the key business processes and the process execution staff. The CBI manager was unable to implement the improvement.

Message: The process execution business managers need to be the 'owners' or process stewards of the business processes. This can never be superseded by the owner of the process modeling tool.

- Version control and release management is critical. This especially applies to situations where BPI projects significantly modify the business processes. Sufficient attention must be given to notifying employees about the changes and providing appropriate training and support. With the increase in on-line business applications there is also an increased possibility for e-learning–based training.
- It is strongly recommended that any formal business process models within the same organization use the same process modeling tool, with the same conventions and process architecture. This will allow the development of an end-to-end business process view that cuts across various organizational functional departments and ensuring re-use.

Process simulation

- *Description*
  This tool will allow the organization to simulate the viability of its redesigned business processes to possibly identify process weak points and process resource bottlenecks. This is where the organization determines if the business process is able to be executed in the expected way. The simulated business process, based on the assumptions made, can evaluate which of the different alternatives is the best prior to making any cost-intensive business process or technology changes. With this in mind, the following fundamental questions should always be asked regarding the functioning of your redesigned business processes. 'Who does what and in which order?' It is not enough to just describe the business processes in order to be able to best judge the dynamic interplay between the different process scenarios.
- *Main benefits*
  - Ability to determine bottlenecks and weaknesses in the process and dependencies of processes and resources, resulting in better results and lower costs. Goldratt's Theory of Constraints (1999) has identified the importance of bottlenecks on the performance of the organization.
  - Ability to compare various processes and scenarios on the basis of their efficiency and speed and shared best practices, which

should result in lower costs and better results. This allows management to perform what-if analysis and validate their assumption.
  - Current process problems can be reviewed through process simulation. This will highlight whether the problem is 'are the assumptions correct' or 'is the process execution (redesigned business process) correct'?

- *Key considerations*
  - Process simulation sounds almost too good to be true, but a word of caution is required. Correct process simulation requires a lot of detailed and correct information. For example, it is important to be able to correctly estimate the total work time that people have available. It is unrealistic to expect that this is 100%, as people have meetings, downtime, exceptions to deal with, etc. Thus the saying 'garbage in–garbage out' is certainly appropriate in this situation, because if the assumptions are erroneous, so will be the results.
  - Data for processes and activities should be obtained as much as possible from the business operations staff, rather than just assumed, estimated or measured using small samples. With an increase in business applications, more information will be available from the business operations and this will allow for the assumptions to be validated.

Process (workflow) engine

- *Description*
  The commonly used term for a process engine is a workflow system, which describes the automation of internal business operations, tasks and transactions that simplify and streamline business processes. A process engine is the software component that executes transactions or events. In order to execute processes via a process engine, the organization must first model their processes either in the process modeling tool provided by the process engine software provider, or in a specialized process modeling tool, described previously. Full automation of the process is often referred to as Straight Through Processing.
- *Main benefits*
  - Ability to automate work that can be standardized, resulting in decreasing cost and throughput time and thereby increasing quality and reliability. Identical work will be processed in an identical way and is not depend on the individual staff interpretation or the channel being used.
  - Ability to route work on the basis of dependencies, skills and availability, resulting in reduced throughput time, better quality and better resource utilization. This provides an overflow mechanism so that in the case of a sudden heavy workload, other people within the organization or a third party will be able to assist. A possibility is to outsource part of the overflow, even to countries with different time zones to ensure that the organization remains on top of any transaction backlog.

- Escalations are identified instantly and routed to the appropriate employees. An audit-trail can be provided so that management has a full understanding of the steps that lead to the escalation and if it was necessary.
- Ability to track and trace specific work, thus being able to provide customers with real-time information about the status of their transaction. Customers become extremely frustrated with organizations that are not able to provide this information. This ability for the customer to enquire on the status of their transactions will significantly reduce the volumes of calls to call centres. Furthermore, it provides statistics on customer behavior and interest.
- Ability of staff to focus on more interesting and important work, resulting with higher staff satisfaction and better quality.
- A process engine can provide improved opportunities for customers to complete self-service activities, as no staff are required to process the work.
- A process engine does not have to be fully automated as it can also be used to guide staff to follow the right steps and ask the right questions. This supported workflow can reduce the requirement for subject matter expertise in the business.
- A process engine allows new products to be brought to the market faster as it may already have links to the underlying business applications.

• *Considerations*

- A process engine has tremendous potential; however, one of the biggest challenges is how to deal with exceptions as exceptions take a significant and disproportional amount of additional software development. Furthermore, in business operations too many exceptions will reduce the benefits of a process engine – not allowing for exceptions will lead to processing errors as people try to 'squeeze' exceptions through. The way how Aveant, in the case study in Chapter 2, dealt with exceptions is very pragmatic. Rather than wasting business application system development time on specifying the requirements for exceptions, they focused on the development of most of the requirements, and added an 'Exception' button on each screen. This allowed the user to circumvent the rigor of the process engine management system and enter the information. A specialist team would review these exceptions and ensured that the exception was a true exception and that it was dealt with properly.
- Making business processes transparent to the customer, through self-service track and trace, requires a significant discipline and maturity level within the organization, as customers may see how a transaction item may stagnate or even be incorrectly completed (e.g. a domestic package delivery that ends up overseas). Thus, the process flow needs to be optimal and the execution needs to be handled well. We recommend that exceptions, such as stagnation or incorrect execution are discovered within the

system, so that it may be corrected prior to the customer noticing it. Call-centre staff will also need to be better trained, as informed customers will no longer allow staff to say – 'your work is in process'.

Business Rules Engine (BRE)

- *Description*:
  This provides *a significant degree* of agility to organizations because it allows an organization to 'extract' business rules from being hard-coded within legacy systems into a BRE. It is part of the drive to give 'power' and flexibility back to the business rather than relying on the ever-present technology bottleneck. BREs today provide the ability for a technically competent business analyst, working within the business (not IT) to change business rules very quickly. Rather than the business specifying business rule changes, giving this to the IT department to review, develop technical specifications, quote, schedule, develop, test and then implement, a business-based business analyst can complete and test the change, thus providing the business with much increased business agility. This ability to provide fast changes must be kept within the bounds of production promotion policies and manage audit needs of the organization, although these policies may require significant review as a result of this new technology.
- *Main benefits*
  - Ability to automate more work, resulting in improved quality and reducing costs and throughput time. BREs ensure that complex calculations are completed consistently and fast.
  - Development time for new applications or maintaining existing application will reduce as the business rules are configurable and do not need to be coded as they are separated from the coding of the program.
  - Ability to test and manage the business rules prior to releasing any changes, resulting in better quality and reduce costs. It allows scenario testing and evaluating the impact on customers and the bottom line.
  - Ability for the business to define, monitor and manage the business rules as they do not have to rely on IT, resulting in more effective, manageable and agile processes.
- *Considerations*
  - BREs allow organizations to be very flexible. However, it is important to maintain control as there may be a maze of business rules that overlap or interrelate. This has been recognized by companies that produce BREs and they are providing more and more tools to monitor and control them.
  - The organization needs to have capable business analysts who understand the business applications and configuration. As the business application needs to have a consistent configuration over time and any additional project has to fit within the existing configuration. An option is to have a few key business analysts in

the Centre of Business Improvement. Some CBIs have an obsession with process models and ignore the BRE. In these situations the process models are characterized as the way the business process 'should' be run, while the BRE defines how the processes 'actually' run.

Customer relationship management (CRM)

- *Description*
  CRM relates to both a management philosophy on how to deal with customers and an application that maintains, supports and manages the relationship of the organization with its customers, including the capture, storage and analysis of customer information. In this context, CRM is an application that has an increasing overlap and integration with BPMS, as BPMS relates more and more to frontline operations and customer self-service. Furthermore, there is an increasing trend towards a more personal approach of customers, requiring more detailed personal information. There are fundamentally three types of CRM applications:

  - Operational – relates to the automation or support of customer processes executed by a sales or service representative from the organization.
  - Collaborative – relates to customer self-service, where the customer has direct communication with the organization without the need to use a representative.
  - Analytical – relates to the analysis of customer data, including transactions and interaction for a variety of purposes.

- *Main benefits*

  - Ability to execute processes either manually or automatically as a result of specific customer characteristics (e.g. overall product portfolio) or history (for example loyalty award for *x* number of years).
  - Reduced costs and higher customer satisfaction due to faster and better customer services as the required information can be more easily located.
  - Details of previous customer interactions will allow for a faster and more effective customer interaction. There is no need for a customer to repeat previous conversations with customer service representatives as this is all recorded. This is especially of importance with the handover from initial sales interactions to services interactions.
  - Ability to up-sell and cross-sell as a result of insight into the customer's history, profile and current portfolio. Service interaction becomes increasingly a vehicle for cross-sell and up-sell.
  - Better marketing, sales and service processes and initiatives as there is better customer information – this relates to both individual customers and market segments as a whole.
  - Assurance that all customer queries, applications and complaints are being addressed, resulting in better customer service and higher sales.

Document management

- *Description*
  Most processes, certainly in the financial services sector, are accompanied by some form of paper. Hence, if an automated BPM solution is implemented without an accompanying integrated document management system (which includes image processing), the organization risks making its business processes extremely fast, and then having to wait for the physical paperwork to catch up. Clearly it is much better, from a business process perspective, to have scanned documents (images) of the paperwork available on an 'as required' basis by a business process. There are organizations that have made a conscious management decision to implement a BPM solution without the document management component. This can place the entire implementation at considerable risk as it may not provide the expected benefits to the business.
- *Main benefits*:
  - Lower costs and better quality in processing documents as the information is readily available.
  - As documents are electronically available, work can be completed from an electronic version, reducing throughput time significantly (no waiting for the paper to arrive). Having documents electronically also means that the processing of those documents can be done anywhere in the world because there is no need for an individual to actually hold the physical paper. In the past much of the delays in processes were caused by transferring hard copy documents from one person to another.
  - Retrieving and tracking documents can be completed more easily, resulting in lower costs and faster throughput times.
- *Considerations*
  - Documents are being used and consulted by a wide range of people for a variety of reasons. It is important to realize this when developing a document system.
  - Legal requirements related to privacy and duration of storage should always be considered.

Integration

- *Description*
  Provides the interface layer between the individual BPMS components and the legacy business applications of the organization. Due to the consolidation in the BPMS vendor arena, the larger survivors have an increasing number of the BPMS components and they position integration as a critical component.
- *Main benefits*:
  - Ability to combine the various BPMS as outlined in this section.
  - Ability to implement BPM automation while keeping the existing business application systems resulting in substantially more benefits with limited costs.

- Ability to reduce redundancy and inconsistencies of data, resulting in reduced costs and improved quality. It is still amazing that there are organizations that require manual updating of customer data in multiple systems all because the systems are not yet integrated.
- Ability to make changes more quickly than is capable via the traditional legacy systems approach, resulting in more agility and substantially lower costs.
- Consistency in the end-to-end business processes because of an integrated process engine. For example, while making a sale, the on-line product portfolio is being used, hence the sales representatives can only sell valid combinations and all the information is available for billing. In the past, several industries, especially telecom, were plagued by sales representatives selling customers combinations of products and components that were technically (or financially) either not optimal or possible. This often only came to light at the time of installation, servicing or billing and caused serious customer service and satisfaction issues. Using an automated product portfolio while making the sale can ensure that only valid product combinations are sold, or that a team of specialists must first review and approve before it can be offered to a customer.

- *Main considerations*:
  - Some projects focus too much on the running of a single application (as that is specified in their scope of work), as a consequence insufficient attention is given on how this application will fit in the overall system architecture and how to run it optimally during operations.
  - Integration can be one of the hardest issues in a project. This needs to be addressed, planned and tested properly to ensure that the applications run well in the overall set of applications.
  - While integrating applications it is important to realize that the data-models, definitions and processes can vary in the different systems. It is crucial to map all this out well in advance, to ensure that there is an unambiguous link between the various applications and indeed an ability to connect at all.
  - While integrating, especially with external parties, security has to be addressed to ensure that only the relevant information is provided to the authorized parties. The security relates to both the communication and the receivers of the information.

Business intelligence (Business Activity Monitoring (BAM))

- *Description*

  This is about the collection and examination of performance-related business process information and it is an essential prerequisite for successfully implementing and evaluating business process performance measures for the continuous optimization of business processes. While we will continue to refer to it here as BAM, it is now often referred to as Business Intelligence by many observers (Davenport and Harris, 2007) and we have outlined this in more detail in Chapter 6 – Process

Performance. BAM provides the actual performance measures that can be compared to the targets set by the organization. It automatically identifies performance data from an organization's business processes perspective, especially those which span business application systems, and thus makes it possible to analyse them. This information can be gathered from the various business application systems within the organization. BAM provides information that helps to uncover weaknesses in business process handling and to optimize processing throughput times. It acts as an early warning system by not only providing historical information, but predictive information, for the monitoring of business processes. Reporting can be via printed reports, or more likely, management dashboards.

- *Main benefits*
  - Ability to monitor business processes in real-time (or near real-time) and drill down into problem areas, resulting in less problems and lower costs.
  - Ability to forecast and identify delays and Service Level Agreements (SLA) that cannot be met and allowing for pro-active mitigation action.
  - Ability to benchmark business processes against competitors and industry standards, resulting in better results.
  - Management has an on-line insight into the progress of the work, the performance of the teams and individuals, and is made aware of processing bottlenecks. Delays due to the absence of staff can be identified and properly addressed quickly.
- *Main considerations*
  - The saying 'garbage in–garbage out' relates particularly to BAM. Meaningful information can only be derived when the underlying data are reliable.
  - It is important to specify what information is required and ensure that people are not flooded with endless reports and graphs. It is important that the right information gets escalated to the right people at the right time.

Activity-based costing (ABC)

- *Description*

  This represents an add-on tool for existing cost accounting systems. ABC makes the success of business operations and BPM projects measurable. It creates transparency in the understanding and control of business process costs. It is a tool to assist in securing strategic organizational decisions on the cost side and to achieve long-term cost reduction. The ability to generate and utilize competitive advantages requires knowledge of the 'right' costs.
- *Main benefits*
  - Ability to understand cost components of processes, resulting in a better alignment between price of products and services and the costs associated with them.

- Ability to compare various business processes and identify areas for improvement, resulting in lower costs.

- *Main considerations*
  - Before applying ABC the value of the information to be obtained should be specified. Too much detailed information will require substantial effort in registering and maintaining cost data, whereas too high level data may lead to oversimplification.

Leveraging service-oriented architecture (SOA)

> To get the best business value out of SOA and BPM, the two initiatives should be aligned at multiple levels, to coordinate and share key resources. This is the most effective way to drive business priorities through the IT infrastructure, and it is a powerful method for aligning business with IT.
>
> Paolo Malinverno, Gartner Group

Definition

SOA provides interesting opportunities to deploy BPM or process-focused 'thinking' into automation. Separating the 'what' from the 'how' has been a lasting problem in the relationship between business and IT – very few business requirements describe clearly the 'what' while avoiding going into detail on the 'how'. The *what* (the service required by the business) should be described in plain English that can be easily comprehended by the business. The *how* (the technical realization of the required service) can be developed by IT without the need for business involvement.

Before progressing any further, we would like to define SOA and unfortunately, like many other three letter acronyms there is no single accepted definition.

Gartner has a rather technical definition of SOA:

> a client/server software design approach in which an application consists of software services and software service consumers (also known as clients or service requesters).

Bieberstein (2005) has a more process-focused definition:

> a framework for integrating business processes and supporting IT infrastructure as secure, standardized components – services that can be reused and combined to address changing business priorities.

Concepts of SOA

Van den Berg, Bieberstein and Van Ommeren (2007) identify the following components of SOA:

1. Componentize – to prevent things from becoming messy, you naturally group things and define components. These components can be easily linked to each other, such as Lego-blocks.

2 Agree on how to do things – this relates to have a common definition and agreement on how to work together.
3 Use what you already have – before buying something new, first see if something you already have fits the need. With the huge investments made in IT so far, re-use should more and more become the norm.
4 From 'made to order' to 'infrastructure' – if there is a fit for purpose ready-made solution on the shelf, it is better to use that than have one built specifically for the business.
5 Facilitate change, continually improve – the only thing that will not change is change itself, so you probably should rely on the fact that, eventually, some things will change – thus flexibility and agility become the norm and need to be built into business applications.
6 Do it for a (business) reason – when spending money, you want to know what you are getting in return. React to the environment. This should be specified upfront and used as a continuous measure.
7 React to the environment – while a bit similar to 'facilitate change', reacting to the environment is the day-to-day situation in which businesses operate – if something happens, organizations need to respond in such a way that is best for the organization; thus close monitoring and management is important.

Introducing SOA can be a major challenge, especially as it is seen as yet another IT gimmick or hype. Van den Berg, Bieberstein and Van Ommeren (2007) propose that the business creates a business vision for SOA to get a commitment from the business. The business vision is a process and it is important to involve key stakeholders and educate and motivate them throughout this process. The five elements of the business vision are:

1 Reasons – the reasons to start with SOA can vary and can be both internal and external to the organization. An example of an internal driver can be the increasing costs of developing an increasing number of interfaces, as the number of internal applications expands. An example of an external driver is the need to provide an interface to a newly established partner. The reason usually has to do with pain or an urgent need.
2 Benefits – determining the SOA benefits is the next step in defining the business vision. It is important to find the most suitable set of benefits. This normally requires several iterations as the participants obtain a better understanding of SOA and its potential. The benefits must be specific enough so that they can be measured. Ideally, these benefits align, just as other projects and business processes, to the overall objectives of the organization.
3 Definition – the business must specify its definition of SOA. A good definition provides all involved with a clear understanding. It is not so important to have a correct definition but rather one that is the most suitable for the organization.
4 Consequences – when deploying SOA the organization will be impacted. It is important to identify all these consequences: impact, desirability and costs. As a result of this analysis an assessment can be made of the overall impact and the cost-benefit.

5 Implementation – needs to be well planned, especially as typically multiple implementations are running in parallel. An implementation strategy outlining the approach is recommended. A key component is an appropriate management structure to oversee the implementation always ensuring that it is in line with the stated benefits.

Common situations

No integration

Often new business applications are selected and implemented in isolation. The features of the product have been seriously and methodologically scrutinized; however, the ability to integrate is unfortunately too often forgotten.

The reasons for these are mainly:

- *Oversight*: The people involved are not aware of the potential and need for integration. The best way to avoid this is to ensure that, at a project level, adequate business analysis is performed and a Project Starting Architecture completed, to ensure that the context is complete.

  At an organizational level it is strongly recommended to have a process architecture approval process (this is usually part of either the Centre of Business Innovation or the Strategic Process Council) that assesses new projects on their adherence to the enterprise architecture as well as their integration with other relevant business applications.
- *Avoidance*: The people are aware of the potential and need for integration, but like to avoid it for as long as possible as it is deemed to complex, time-consuming and expensive. This is often assessed from a limited project view to meet project timelines and project budgets rather than a long-term sustainable perspective.

  At a project level the best way to deal with this is to include the full-costing, not just for the project but also for the on-going support in business as usual, including the consequences of the non-integration.

  At an organizational level the process architecture approval process should have the power to stop or modify proposals and projects that do not adhere to the organizations architectures.

No business requirements

Recently we have often heard that business requirements documentation is not required anymore as they are old-fashioned. This results in the project becoming technology-centric and the outcome of the pilot will drive the project outcome. This is a dangerous path as there are no formal documents which describe the requirements from the business.

We strongly recommend the definition of business requirements, no matter what type of technology is used. This document must at least describe the:

- Purpose of the project.
- Objectives, deliverables and outcomes of the project.

- Scope (including a listing of out of scope). This should include the Red Wine Test outcomes.
- Project dependencies and implementation considerations.
- Analysis of the current situation.
- Outline of the proposed solution, including processes, and the impact on the organization and business.

Bibliography

Align Journal (2007) *Alignment of Governance: Solving the Critical Linkages Between Corporate and IT Governance*, May/June.

American Productivity and Quality Center (2005) *Air Products and Chemicals Inc. Case Study*.

Bieberstein, N., Bose, S., Walker, L. and Lynch, A. (2005) *Impact of service-oriented architecture on enterprise systems, organizational structures, and individuals*.

Bossidy, L., Charan, R. and Burck, C. (2002) *Execution: The Discipline of Getting Things Done*, Crown Publishing Group.

BPTrends Survey (2007) *A Survey of Business Process Initiatives*, authored by Nathaniel Palmer and published on BPTrends.com in January.

CRM Today (2006) *Business Intelligence Software Market to Reach $3 Billion by 2009*, http://www.crm2day.com/news/crm/117297.php

Davenport, T. (2001) *The Attention Economy: Understanding the New Currency of Business*, Harvard Business School Press.

Davenport, T. and Harris, J. (2007) *Competing on Analytics, the New Science of Winning*, Harvard Business School Press.

De Waal, A.A. (2005) *The Foundations of Nirvana*, Unpublished, May.

Dundon, E. (2002) *Seeds of Innovation, Cultivating the Synergy that Fosters New Ideas*, Amacon.

Eyck van Heslinga, H.C. van. (2002) *Hands-on crisis management*, Kluwer, Deventer.

Fingar, P. (2006) *Extreme Competition: Innovation and the Great 21st Century Business Reformation*, Meghan Kiffer.

Fingar, P. (2007) *Extreme Competition: Shift Happens*, BP Trends, 01 May.

Gernster, L.V.Jr. (2002) *Who Says Elephants Can't Dance?*, Harper Business.

Goldratt, L. and Eliyahu, M. (1999) *Theory of Constraints*, North River Press.

Goodpasture, J.C. (2000) *Proceedings of the Project Management Institute Annual Seminars & Symposium*, September 7–16, Houston, Texas, USA.

Google (2007) Google.com, accessed 12 November 2007.

Hammer, M. (1993) *Fortune*, Vol. 4, October.

Hammer, M. (1994) *Fortune*, Vol. 22, August.

Hammer, M. (2004) How operational innovation can transform your company, April, Harvard Business Review.

Hammer, M. and Champy, J. (1990) Reegineering work: don't automate, obliterate. *Harvard Business Review*, July.

Hardjono, T.W. (2007) White Paper: *Maturity of Organizations and Business Excellence—The Four-Phase Model.*
Harmon, P. (2003) *Business Process Architecture and the Process-Centric Company*, Business Process Trends.
Hawley, J. (1993) *Reawakening the Spirit in Work*, Berrett-Koehler Publishers.
Hersey, P., Blanchard, K. and Johnson, D. (2005) Management of Organizational Behavior: Leading Human Resources (8th Edition).
Jeston, J. and Nelis, J. (2008) *Business Process Management: Practical Guidelines to Successful Implementations*, Butterworth-Heinemann.
Johan (2007) http://www.gordontraining.com/article-leaving-a-new-still-is-easier-said-than-done.html, accessed 23 October 2007.
Kaplan, R.S. and Norton, P.D. (2008) *Mastering the Management System*, Harvard Business Review, January.
Kendall, G. (2003) *Project Management*, unpublished.
Kendall, G. *Critical Chain and the PMO*, unpublished.
Kepner-Tregoe, Inc. (1995) *Research Report on People and their Jobs: What's Real, What's Rhetoric?*, Internally publication.
Kim, W.C. and Mauborgne, R. (2005) *Blue Ocean Strategy: How to Create Uncontested Market Space and Make the Competition Irrelevant*, Harvard Business School Press.
Kleiner, A. (2005) Beware the product death cycle. *Strategy + Business,* 38, Spring.
Kocourek, P., Newfrock, J. and Van Lee, R. (2005) SOX rocks, but won't block shocks. *Strategy + Business,* 38, Spring.
KPMG (1999) Mergers and acquisitions global research report, KPMG London.
Krames, J. (2003) *What the Best CEOs Know: 7 Exceptional Leaders and Their Lessons for Transforming any Business*, McGraw Hill.
The McKinsey Quarterly (2007) *Better strategy through organizational design*, by L.L. Bryan and C.I. Joyce, Number 2.
Mintzberg (2005) *Strategy Safari: A Guided Tour Through The Wilds of Strategic Management* (Paperback), Palladium (2007) *Strategy Focussed Organization Assessment*, a 360o evaluation of Strategy Execution.
Montgomery, C.A. (2008) *Putting the Leadership Back into Strategy,* Harvard Business Review, January.
Office of Government Commerce (2002) *Managing Successful Projects with PRINCE2* Third Edition, published for the Office of Government Commerce (UK) under licence from the Controller of Her Majesty's Stationery Office.
Office of Government Commerce (2006) *Portfolio, Programme & Project Management Maturity Model* (P3M3), Version 1.0.
Malinverno, P. and Gartner Group (2007) *Gartner Application Architecture, Development & Integration Summit Nashville*, TN Gaylord Opryland Resort & Convention Center, 11–13 June.
Porter, M. (1980) *Competitive Strategy: Techniques for Analyzing Industries and Competitors*, Free Press.
Rohm, H. and H. Larry (2005) White paper: *Developing and Using Balanced Scorecard Performance Systems.*
SAS Institute Nederland (2002) *Nederlandse topmanagers missen informatie uit eigen bedrijf bij strategische beslissingen*, SAS, Nederland.
Smith, H. and Fingar, P. (2004) *Business Process Management – The Third Wave*, Meghan-Kiffer Press.
Souder, W.E. (1988) In Katz, R. (eds), *Managing Professionals in Innovative Organisation*, Ballinger (1988).
Spanyi, A. (2003) *Business Process Management is a Team Sport: Play it to Win!*, Anclote Press.
Taylor, F.W. (1998) *The Principles of Scientific Management*, Dover Publications. (reprint of 1911 original).
Tijdschrift Controlling (2005) *Slechts tweederde directeuren zeker van baan.* Research Nederlands Centrum van Directeuren en Commissarissen; Rost van Tonningen, B. (2002), Wat is anno 2002 de Boardroom Agenda, *Management.*
Van den Berg, M. and Steenbergen, M. (2006) Building an Enterprise *Architecture Practice, Tools, tips, best-practices*, ready-to-use insights.
Van den Berg, M., Bieberstein, N. and van Ommeren, E. (2007) *SOA for Profit, A Manager's Guide to Success with Service Oriented Architecture*, Sogeti & IBM.
Wagter, R., van den Berg, M., Luijpers, J. and van Steenbergen, M. (2005) *Dynamic Enterprise Architecture: How to Make IT Work*, John Wiley & Sons.
Walton, M. (1986) *The Deming Management Methods,* The Berkley Publishing Group.

Index